Systems Microbiology

Current Topics and Applications

Edited by

Brian D. Robertson

Centre for Integrated Systems Biology and Bioinformatics
Imperial College
London

and

Brendan W. Wren

London School of Hygiene and Tropical Medicine
London
UK

Caister Academic Press

Caister Academic Press
Norfolk, UK

www.caister.com

British Library Cataloguing-in-Publication Data
A catalogue record for this book is available from the British Library

ISBN: 978-1-908230-02-7

Cover design adapted from Figures 3.2 and 6.3.

Printed and bound in Great Britain

Contents

Contributors

Eivind Almaas
Department of Biotechnology
Norwegian University of Science and Technology
Trondheim
Norway

eivind.almaas@ntnu.no

Judith P. Armitage
Department of Biochemistry and Oxford Centre for Integrative Systems Biology
University of Oxford
Oxford
UK

judith.armitage@bioch.ox.ac.uk

Chris P. Barnes
Centre for Bioinformatics
Division of Molecular Biosciences
Imperial College London
London
UK

christopher.barnes@imperial.ac.uk

Rico Barsacchi
Max Planck Institute for Molecular Cell Biology and Genetics
Dresden
Germany

barsacch@mpi-cbg.de

Dany J.V. Beste
Faculty of Health and Medical Sciences
University of Surrey
Guildford
UK

d.beste@surrey.ac.uk

Per Bruheim
Department of Biotechnology
Norwegian University of Science and Technology
Trondheim
Norway

bruheim@biotech.ntnu.no

Dirk Bumann
Biozentrum
University of Basel
Basel
Switzerland

dirk.bumann@unibas.ch

Diana Clausznitzer
Division of Molecular Biosciences and Centre for Integrative Systems Biology
Imperial College
London
UK

diana.clausznitzer@bioquant.uni-heidelberg.de

Robert G. Endres
Division of Molecular Biosciences & Centre for Integrative Systems Biology
Imperial College
London
UK

r.endres@imperial.ac.uk

Navin Gopaldass
Departement de Biochimie
Faculte des Sciences
Universite de Geneve
Geneva
Switzerland

navin.gopaldass.unige.ch

Nathan Harmston
Centre for Bioinformatics
Division of Molecular Biosciences
Imperial College London
London
UK

nathan.harmston07@imperial.ac.uk

Maxime Huvet
Centre for Bioinformatics
Division of Molecular Biosciences
Imperial College London
London
UK

m.huvet@imperial.ac.uk

Alexey Kolodkin
Netherlands Institute for Systems Biology
VU University Amsterdam
Amsterdam
The Netherlands;
Luxembourg Centre for Systems Biomedicine
University of Luxembourg
Esch-sur-Alzette
Luxembourg

alexey.kodolkin@uni.lu

Kees Korbee
Department of Infectious Diseases
Leiden University Medical Center
Leiden
The Netherlands

c.j.korbee@lumc.nl

Theresa Kouril
Faculty of Chemistry
Biofilm Centre
Molecular Enzymetechnology and Biochemistry
University of Duisburg-Essen
Essen
Germany

theresa.kouril@uni-due.de

Rahmi Lale
Department of Biotechnology
Norwegian University of Science and Technology
Trondheim
Norway

rahmi.lale@biotech.ntnu.no

Johnjoe McFadden
Faculty of Health and Medical Sciences
Guildford
UK

j.mcfadden@surrey.ac.uk

Jacques Neefjes
The Netherlands Cancer Institute
Amsterdam
The Netherlands

j.neefjes@nki.nl

Tom H.M. Ottenhoff
Department of Infectious Diseases
Leiden University Medical Center
Leiden
The Netherlands

t.h.m.ottenhoff@lumc.nl

Brian D. Robertson
Centre for Integrated Systems Biology and Bioinformatics
Imperial College
London
UK

b.robertson@imperial.ac.uk

Peter Ruoff
Centre of Organelle Research
University of Stavanger
Stavanger
Norway

ruoff@ux.uis.no

Tiziana Scanu
The Netherlands Cancer Institute
Amsterdam
The Netherlands

t.scanu.nki.nl

Bettina Siebers
Faculty of Chemistry
Biofilm Centre
Molecular Enzymetechnology and Biochemistry
University of Duisburg-Essen
Essen
Germany

bettina.siebers@uni-due.de

Jacky Snoep
Netherlands Institute for Systems Biology
Free University Amsterdam
Amsterdam
The Netherlands
and
Manchester Centre for Integrative Systems Biology
University of Manchester
Manchester
UK
and
Department of Biochemistry
Stellenbosch University
Matieland
South Africa

j.l.snoep@vu.nl

Thierry Soldati
Departement de Biochimie
Faculte des Sciences
Universite de Geneve
Geneva
Switzerland

thierry.soldati@unige.ch

Ralf Steuer
Manchester Centre for Integrative Systems Biology
University of Manchester
Manchester
UK

steuer.ralf@googlemail.com

Michael P.H. Stumpf
Centre for Bioinformatics
Division of Molecular Biosciences
Imperial College London
London
UK

m.stumpf@imperial.ac.uk

Varadharajan Sundaramurthy
Max Planck Institute for Molecular Cell Biology and Genetics
Dresden
Germany

sundaram@mpi-cbg.de

Sylvain Tollis
Division of Molecular Biosciences & Centre for Integrative Systems Biology
Imperial College
London
UK

s.tollis@iecb.u-bordeaux.fr

Svein Valla
Department of Biotechnology
Norwegian University of Science and Technology
Trondheim
Norway

svein.valla@biotech.ntnu.no

Hans V. Westerhoff
Netherlands Institute for Systems Biology
Free University Amsterdam
Amsterdam
The Netherlands;
Manchester Centre for Integrative Systems Biology
University of Manchester
Manchester
UK

hans.westerhoff@manchester.ac.uk

Brendan Wren
London School of Hygiene and Tropical Medicine and Centre for Integrated Systems Biology and Bioinformatics
Imperial College
London
UK

brendan.wren@lshtm.ac.uk

Melanie Zaparty
Institute for Molecular and Cellular Anatomy
University of Regensburg
Regensburg
Germany

melanie.zaparty@vkl.uni-regensburg.de

Marino Zerial
Max Planck Institute for Molecular Cell Biology and Genetics
Dresden
Germany

zerial@mpi-cbg.de

Dedicated to our deceased colleagues, Jaroslav Stark and Emmanuelle Caron, pioneers in systems biology.

Current Books of Interest

Horizontal Gene Transfer in Microorganisms	2012
Microbial Ecological Theory: Current Perspectives	2012
Two-Component Systems in Bacteria	2012
Foodborne and Waterborne Bacterial Pathogens	2012
Yersinia: Systems Biology and Control	2012
Stress Response in Microbiology	2012
Bacterial Regulatory Networks	2012
Systems Microbiology: Current Topics and Applications	2012
Quantitative Real-time PCR in Applied Microbiology	2012
Bacterial Spores: Current Research and Applications	2012
Small DNA Tumour Viruses	2012
Extremophiles: Microbiology and Biotechnology	2012
Bacillus: Cellular and Molecular Biology	2012
Microbial Biofilms: Current Research and Applications	2012
Bacterial Glycomics: Current Research, Technology and Applications	2012
Non-coding RNAs and Epigenetic Regulation of Gene Expression	2012
Brucella: Molecular Microbiology and Genomics	2012
Molecular Virology and Control of Flaviviruses	2012
Bacterial Pathogenesis: Molecular and Cellular Mechanisms	2012
Bunyaviridae: Molecular and Cellular Biology	2011
Emerging Trends in Antibacterial Discovery: Answering the Call to Arms	2011
Epigenetics: A Reference Manual	2011
Metagenomics: Current Innovations and Future Trends	2011
Nitrogen Cycling in Bacteria: Molecular Analysis	2011
The Biology of Paramyxoviruses	2011
Helicobacter pylori	2011
Microbial Bioremediation of Non-metals: Current Research	2011
Lactic Acid Bacteria and Bifidobacteria: Current Progress in Advanced Research	2011
Viruses and Interferon: Current Research	2011
Essentials of Veterinary Parasitology	2011
Hepatitis C: Antiviral Drug Discovery and Development	2011
Streptomyces: Molecular Biology and Biotechnology	2011
Alphaherpesviruses: Molecular Virology	2011
Recent Advances in Plant Virology	2011
Vaccine Design: Innovative Approaches and Novel Strategies	2011
Salmonella: From Genome to Function	2011
PCR Troubleshooting and Optimization: The Essential Guide	2011
Insect Virology	2010
Environmental Microbiology: Current Technology and Water Applications	2011
Sensory Mechanisms in Bacteria: Molecular Aspects of Signal Recognition	2010
Bifidobacteria: Genomics and Molecular Aspects	2010

Preface

Systems biology aims to study the dynamic interactions of more than one component in a biological system in order to understand and predict the behaviour of the system as a whole. Typical approaches involve an iterative cycle of 'dry lab' modelling and 'wet lab' verification.

Bringing order to biological data that inherently have noise due to a multitude of variables has previously been considered too challenging. However, systems biology is now a rapidly expanding discipline fuelled by the 'omics' era that is coupled to several new technological advances that have increased the precision of data obtainable. This has provided the bit parts of complex living cells. New challenges arise to put these levels of information together including finding a common language of the difference omics data sets (e.g. genomics, transcriptomics, proteomics, metabolomics).

The sheer complexity of biological systems means that systems biology is a fledgling science. However, a focus on simple single cell organisms such as bacteria aids tractability and means that systems microbiology is a rapidly maturing science.

This book will include case studies on single microbial species (e.g. bacteria and archaea), systems analysis of microbial phenomena (e.g. chemotaxis and phagocytosis) and this is complemented with theoretical approaches and mathematical modelling.

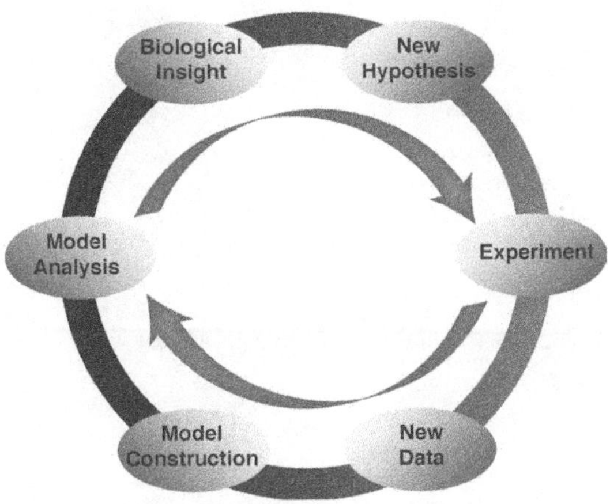

Figure 0.1 Systems biology illustrating the iterative cycle of 'dry lab' modelling and 'wet lab' verification. Concept and design by Jaroslav Stark.

Mathematical Models for Systems Biology and How to Construct Them

1

Chris P. Barnes, Maxime Huvet, Nathan Harmston and Michael P.H. Stumpf

Abstract

Modelling methodologies in the life and biomedical sciences are hampered by the complexity of the processes and systems at work. Modelling studies into prokaryotic systems require the elucidation of the mechanistic model. In this chapter we introduction modelling methodologies and discuss the problem of model (and parameter) inference. We comment on state-of-the-art research questions and provide a general discussion on how models can and should be used in order to better understand the structure, function and dynamics of biological systems. The aim is not to provide an introduction to modelling per se, but to provide readers with an overview on the available methodologies. The modelling approach chosen depends on the biological question at hand as well as a range of other factors.

Introduction

Biological systems work exceedingly well – of that there can be no doubt – and most of them can cope with a bewildering range of different environmental circumstances (Kussel and Leibler, 2005; Nurse and Hayles, 2011). Feats of human engineering often fail to impress in comparison to what comes so naturally to biological organisms. However, with perhaps a few exceptions, we do understand really rather well how engineered systems operate and we have established and amply proven frameworks that allow us to design and construct engineered systems that can fulfil defined functions (Stricker *et al.*, 2008). Often the means by which this is possible rely on our ability to abstract processes (ranging from the transmission of information using electromagnetic radiation to keeping planes in the air) in concise mathematical form. These models are typically constructed as convenient representations of reality, or at least those aspects of reality believed or known to be important in a given situation (May, 2004).

Progress in the physical and engineering sciences has rested to a large extent on the tight integration between experiment and (mathematical) theory. Recent years have seen a gulf opening up between theoretical physicists, i.e. researchers using mathematical tools to study physical processes, and mathematical physicists, i.e. those who study physics-inspired mathematical objects and theories. The former perspective is successful in parts because there is a canon of models that are familiar to every physicist, and which serve as common reference points that enable conversation and collaboration between experimental and theoretical researchers. In the biological domain such widespread acceptance of models and quantitative approaches is still outstanding. This is despite the fact that the use of mathematics in biology predates the disciplined use of mathematics in the physical sciences, for Leonardo de Pisa, otherwise better known as Fibonacci, modelled rabbit population dynamics some 200 years before Galileo Galilei established his principles of inertia and laid the foundations for modern physics more generally.

One of the reasons for the slow acceptance of modelling methodologies in the life and biomedical sciences is the complexity of the processes and systems at work. Any attempt at abstracting the structure and dynamics of some systems in terms of mathematical equations may seem hopelessly naive. Given the wealth of data that have been collected for different systems it may also be hard to come up with models that are in agreement with each and every experimental observation. And finally, the number of potential mechanisms that could underlie processes such as the transcription and translation of a gene into its protein products, signal transduction, endocytosis, cell-to-cell communication or control of the cell cycle is vast. This in many ways turns out to be the main problem for modelling of biological systems: while systems biology is data rich, the field is even richer in hypotheses. Here we may identify different mathematical models with each hypothesis, and choosing which one to pursue is a difficult choice. So difficult in fact, that once the nature of the model is specified, any further mathematical analysis is straightforward by comparison.

Thus for any modelling study into prokaryotic (or any other biological) system, elucidation of the mechanistic model is the real challenge. We therefore emphasize this aspect in particular in this chapter. After a brief introduction to modelling methodologies we turn our attention to the problem of model (and parameter) inference. Here we attempt to provide some guidance as to how best to proceed. This touches upon state-of-the-art research questions in statistics (Cox, 2006) and machine learning (Hastie *et al.*, 2003), but as much as possible we have tried to give a balanced overview. We conclude the chapter with a more general discussion as to how models can and should be used in order to better understand the structure, function and dynamics of biological systems. The aim is not to provide an introduction to modelling per se, but to provide readers with an overview over the available methodologies. Which modelling approach is then ultimately chosen depends on the biological question at hand, but also potentially on a host of social factors, for example which modelling approach is favoured by an available and interested colleague and potential collaborator.

The use of models

Framing biological problems in terms of mathematical models forces us to explicitly state our assumptions about how biological processes occur and how systems operate. Rarely if ever do we have complete knowledge of all the interactions and their determinants (e.g. whether a physical protein–protein interactions is realized during a particular phase of the cell cycle; Kelly and Stumpf, 2008). In mathematical models we are not limited to capturing what is known, but are also free to explore new or alternative mechanistic hypotheses. Once a model has been constructed – and how this can be achieved is discussed below – we use mathematical or computational approaches in order to arrive at predictions as to how these alternative models might differ, or what unobserved data might be expected under suitable conditions. The mathematical analysis is simply a rigorous way to arrive at such predictions for the behaviour of complex systems.

In addition to understanding the molecular mechanisms underlying biological systems

– especially those not directly observable in experiments – mathematical models can also be used to predict the behaviour of a system. This may, for example, be useful to plan therapeutic interventions. Interestingly we often observe a trade-off between a model's explanatory usefulness and its predictive power (Hastie *et al.*, 2003). Simple models can do an outstanding job for predictions, even though they capture only a small amount of what is known to be important for understanding the inner workings of a system. In capturing less detail predictive models may become less biased and thereby gain in predictive power. This is, of course, always conditional on the right aspects of a system being included into the predictive model.

While the mathematical analysis of models can involve highly involved and sophisticated techniques, good models can be illuminating even to those with little expertise. In particular in suitable graphical representations they can be useful tools that facilitate the interactions between experimental and clinical scientists, on the one hand, and theoretical researchers, on the other. Such dialogue is, of course, pivotal for the development of models.

Mathematical modelling approaches

How to model a system depends on a number of factors: first and foremost, the choice of modelling framework should depend on the problem and available data at hand (Fig. 1.1). There is no point in using highly sophisticated computational techniques and detailed mathematical models of dynamical systems, if the only data available are knock-out phenotypes. Similarly, if we have access to temporally highly resolved data, then static-graph based or simple logical descriptors will only offer limited insights. A degree of flexibility is therefore

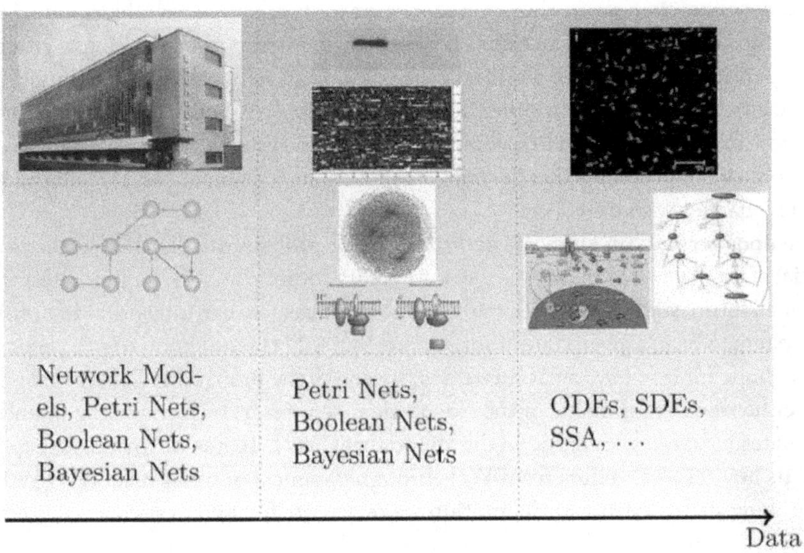

Figure 1.1 The choice of modelling techniques depends on the amount, quality and resolution of the experimental data. As more and better resolved data are becoming available, modelling approaches can or need to become more sophisticated. At the same time it becomes harder to develop good models and the cost of statistical analysis increases.

required from the modeller, and insistence on a single modelling framework for every given model should be considered a warning sign that engagement with the biological problem will be superficial.

Naturally, the more detailed the data are, the richer a mathematical formalism will typically be required. The simplest way of generating a model is simply to draw graphs (or networks) that link the essential components of a biological systems, displaying either interactions or statistical dependencies (Alm and Arkin, 2003; de Silva and Stumpf, 2005). Here we use a description in terms of graphs,

$$G = (N, \varepsilon) \tag{1.1}$$

where N is the set of nodes (e.g. genes or proteins) and ε the set of interactions among them. Such network-based summaries allow us to display in concise form the sets of relationships known, believed or postulated to exist. In principle they do not carry more information about a system than a list or table of interactions would, but, provided that the network is not too large, network visualization can aid our understanding of complex systems in a way lists cannot.

It is naive to believe that static graphs provide an adequate description for real biological systems, as networks of physical and regulatory interactions will change with time and in response to environmental, developmental and physiological stimuli (Kelly and Stumpf, 2008; Lèbre *et al.*, 2010). Thus, networks change over time all the time and present protein-protein, metabolic and gene regulatory networks are only a depiction of interactions for which there is some experimental – though potentially erroneous – evidence that they can occur.

Building on these static descriptions, however, we can develop dynamical descriptions. Here we describe changes in the abundances of molecular species, the nodes, N, in the network, through sets of interactions, ε, which may correspond to chemical reactions or regulatory relationships. In the simplest case of Boolean networks (Bornholdt, 2008), the nodes are either 'on' or 'off', but other descriptions exist (e.g. Petri and Bayesian networks) which allow for more nuanced behaviour. In all of these systems we also have mathematical descriptions of the processes that describe how the states/abundances, $Y(\mathrm{t})$, of the molecular species changes with time.

These update rules can either be deterministic or probabilistic. In the former case, once the initial state $(Y(t=0))$ of a system is known, its future state can be predicted with certainty for all future states ($Y(t)$ with $t>0$); in the latter case we can only provide a probability for the system to be in a given state at some time, that is $Y(t)$ is not known with certainty, and all we can hope for is to have an accurate description of the probability $\Pr(Y(t))$ for all times $t \geq 0$. In either the deterministic or the stochastic case we start by formulating mathematical models in terms of rates of change in the abundances, $Y(t)$. That is we formulate expressions that tell us how $Y(t+\tau)$ differs from $Y(t)$ through chemical or other reactions (Wilkinson, 2006). Taking slight liberties with notation we may write for such systems:

$$\frac{dY(t)}{dt} = f\left(Y(tt),\ ;\theta\right) \tag{1.2}$$

where θ is the vector of parameters describing the temporal evolution of $Y(t)$ (the liberty we have taken here is to subsume any stochastic effects into the model f(...) rather than making them explicit); here we always assume that the initial state $Y(0)$ is known or can be controlled. The model $f(Y(t),t;\theta)$ includes the 'network structure' as well as other factors such as any stoichiometric considerations.

A simple example of such a model described the translation and transcription of DNA into protein. The simplest such model has the form

$$\begin{array}{ccccc} \emptyset & \xrightarrow{k_m} & \text{mRNA} & \xrightarrow{k_p} & \text{Protein} \\ & & \gamma_m \big\downarrow & & \gamma_p \big\downarrow \\ & & \emptyset & & \emptyset \end{array} \tag{1.3}$$

Here, as nearly always, a simple graphical depiction of the model hones intuition and can serve as a means of communicating ideas between experimentalists and modellers. Similarly, graphical notation can simplify the development of mathematical models. Here the $\emptyset$ at the left denotes that we assume that mRNA is produced de novo once a certain stimulus has been triggered. Biologically that corresponds, of course, to a transcription factor binding event, or a similar trigger of transcription. The rates k_m and k_p are the transcription and translation rates, respectively, and γ_m and γ_p denote degradation of mRNA and proteins.

In mathematical form we can write this model in its deterministic form as (Elowitz *et al.*, 2002; Ingram *et al.*, 2008):

$$\frac{d[\text{mRNA}]}{dt} = \kappa_r - \gamma_r[\text{mRNA}] \tag{1.4}$$

$$\frac{d[\text{Protein}]}{dt} = \kappa_p[\text{mRNA}] - \gamma_p[\text{Protein}] \tag{1.5}$$

In Fig. 1.2 we compare and contrast deterministic and stochastic simulations for this model. While for the deterministic model the system approaches a finite steady state very quickly, the stochastic dynamics exhibit fluctuations over a wide range of mRNA and protein abundances. Averaging over many stochastic simulations, however, yields the same behaviour as the corresponding deterministic model. This is, however, not necessarily so. First of all, stochasticity will frequently result in extinctions of some molecular species through chance alone; second, the fluctuations can be so pronounced to overwhelm the average behaviour, such that no single simulation adheres to the mean behaviour. In gene expression for example we may speak of genes being turned 'ON' or 'OFF' and observe that for example the gene is in the 'ON' state in 30% of cells. If we measure the average expression level this could suggest that the activity of the gene amounts to 30% of the maximum activity level, which is, however, meaningless; finally, for non-linear problems the mean behaviour can also be different from the deterministic behaviour under fairly general conditions (Erguler and Stumpf, 2008).

Also shown in the same are trajectories for the repressilator (Elowitz and Leibler, 2000), one of the canonical models for synthetic biological systems. Both the deterministic and stochastic system exhibit oscillatory behaviour in the proteins making up these systems, but

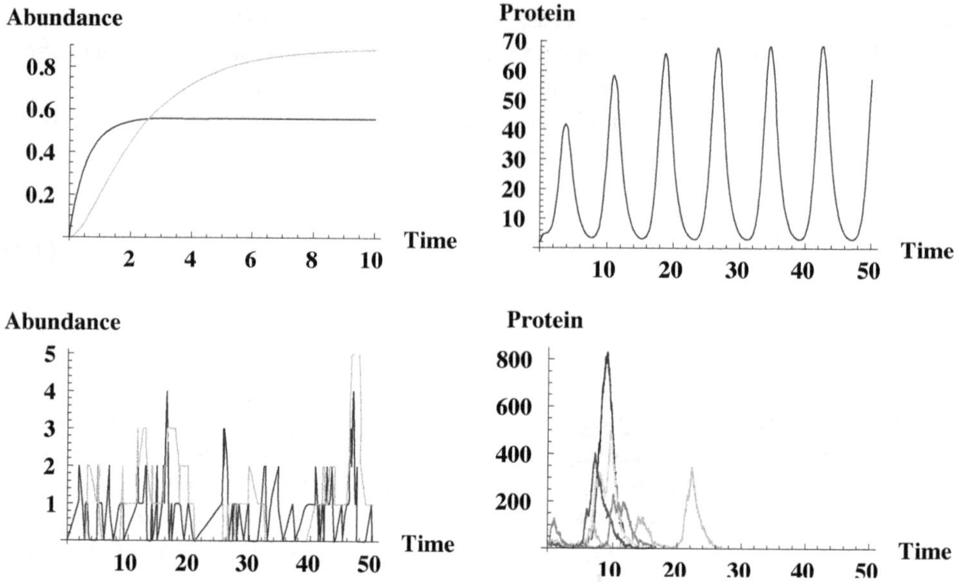

Figure 1.2 Deterministic (top left) and stochastic simulations (bottom left) of the simple model of gene expression. The total abundances of mRNA and protein behave very differently for the two modelling frameworks. While the deterministic dynamics result in stable equilibrium states, the stochastic dynamics exhibit high levels of variability, which is not observable from the average behaviour exhibited by populations of many cells. The panels on the right of the same figure show the concentrations of one protein in the repressilator model (parameters were taken from Toni *et al.*, 2009). Both the deterministic (top right) and stochastic (bottom right) models show oscillatory behaviour, but the five independent runs of the stochastic repressilator model demonstrate the likely differences exhibited in different realizations of the stochastic repressilator.

the different realisations of the stochastic repressilator exhibit considerable variability. This reflects the interplay between the underlying (deterministic) dynamics and the stochastic effects resulting from noise and the small number of many molecular species in the cellular interior.

The choice of stochastic versus deterministic modelling is only one of a potentially bewildering number of choices (Wilkinson, 2009) we have to make when setting out to model a biological system. There are, however, many other choices as well, which depend sometimes subtly and nearly always subjectively, on the problem at hand. Whether to include spatial aspects of a system is a question of growing importance in many areas of systems biology (Kholodenko *et al.*, 2010). Related to this is the choice of whether we should model systems in a continuous (in terms of space, time and state) or granular fashion. Here our aim cannot be to attempt an overview over the wealth of modelling approaches that have been proposed and are being used by the modelling community in systems biology. Several excellent reviews exist already (Bornholdt, 2005; Alon, 2007; Gunawardena, 2010; Kholodenko *et al.*, 2010; Wilkinson, 2009), and new modelling approaches are entering the fray from other disciplines, in particular the physical and engineering sciences. Rather in what follows we will attempt to give an overview over the different approaches that exist for finding suitable models in the first place.

Reverse engineering of biological models

In systems biology models are still hard to come by for most problems, but the fundamental statistical principles are well developed and understood. Here we give a flavour of the different approaches that can be employed in order to learn the structure and dynamics of mathematical models describing biological systems (Stigler *et al.*, 2007). In biology we have, broadly speaking, three avenues to explore:

- Data analysis: performing experiments coupled to statistical inference allows us to calibrate models against data, or even infer model structures.
- Comparative approaches: The single common origin of all life on earth allows us to transfer – with some caution and due diligence – insights gained in one model organism to other systems.
- Text mining: published research is now increasingly searchable electronically, which allows us to cross-link and relate different pieces of information.

All three avenues are worthy of exploration and have become important research areas in their own right. One of the challenges for the future will be the integration of these three information sources into a mutually informative inferential framework. Below we outline these complementary methods, before concluding with a discussion of the current state of the reverse engineering field.

Statistical inference for systems biology

Statistical inference concerns the relation of data to mathematical models in order to make predictions or to understand the underlying physical process. This can be subdivided into two main questions that we would like to answer in the light of some observed data; what does the data inform me about the parameters in my mathematical model? which model is the most appropriate to explain the observations? The first question is referred to as parameter inference, parameter estimation or model calibration while the second is known as hypothesis testing, model comparison or model selection (Cox, 2006). Often these two procedures are combined; a hypothesis test often relies on the estimation of an underlying parameter from the data.

In systems biology we often want to try to relate models to data such that we can make predictions in the case of a new perturbation or other experimental conditions that we would like to test experimentally. This requires creating models that attempt to accurately describe the underlying physical processes such as a signalling cascade or a genetic network. Often these models will describe biochemical reactions using differential equations though other formalisms exist such as Boolean networks. This is in contrast to traditional statistical approaches that aim to describe data using a model that captures the essential features of the data without modelling the underlying process. An example of which would be fitting a regression line to some data. Although these types of models do have uses in systems biology they are far less common than mechanistic models as they rarely give us information on the underlying physical process (though may provide accurate predictions) (Toni *et al.*, 2011).

Mechanistic models often contain many more parameters than descriptive models, many of which cannot be directly measured in the living system. In these situations the parameter values must be inferred from data. Additionally, current experimental methods can usually interrogate only a small subset of all the species of the system in any one experiment. The system often contains many more additional species that remain unobserved or 'hidden'. Since mechanistic models usually contain a high degree of complexity to reflect the underlying biological process, there is often a set of models that are consistent with the data (Stigler *et al.*, 2007). These aspects make statistical inference for systems biology challenging and we generally require sophisticated methods to make reliable inferences and to extract the most information from the data. In this regard, the Bayesian approach to statistical inference is a consistent and powerful framework in order to relate our models to the observed data.

Traditional approaches to parameter estimation and hypothesis testing

The likelihood mathematically expresses the relationship between our model and the observed data. In the most frequent framework it is viewed as a function of the parameters in the model. As an example, when fitting a normal distribution to some height data, represented by $\{x_1, x_2, \ldots, x_N\}$, the likelihood as a function of the mean and variance (denoted by μ and σ^2) is the product of the probability density of each point [in this case the product of normal densities $N(x_i; \mu,\sigma^2)$]. The principle of maximum likelihood states that the most suitable parameters are the ones for which the likelihood is maximal. Thus parameter estimation can be viewed as a non-linear optimization (or minimization) problem with the objective function given by $L(\theta)$ [or $-L(\theta)$]. Often in physics and engineering the explicit dependence on the likelihood is dropped and the optimization is done with respect to a cost function (though the choice of cost function is equivalent to a statement about the likelihood). In our simple example the parameter values that maximize the likelihood are given by mean(x) and Var(x) for μ and σ^2, respectively, which explicitly demonstrates the normal assumption when using these summary statistics. In practice we can rarely derive the maximal values analytically and we must resort to optimization algorithms that maximize the likelihood numerically. Associated with parameter estimates are confidence intervals. These specify regions in which the parameter estimates are most likely to lie. For example a 95% confidence limit on the estimate of the mean in the normal model is given by mean $(x) \pm 1.96\sqrt{Var(x)}$. Again for realistic models we must estimate these regions numerically using asymptotic approximations (which assume the likelihood is normal about its maximal value) or profile-likelihood methods that follow the contours of the likelihood function (Cox and Hinkley, 1974).

We cannot in general compare models through their likelihood functions since they are only defined up to multiplicative constant (only the maximal value is important) and a more complex model (larger number of parameters) will always fit the data better than a simpler model (have a higher likelihood). The question is whether the more complex model is truly more predictive as opposed to just over fitting the observed data. Therefore model selection in the most frequent framework aims to strike a balance between the increased likelihood a more complex model achieves and the extra parameters it introduces. As such there are different methods for model comparison depending on the nature of the models under consideration. The most familiar is traditional hypothesis testing that concerns the comparison of models that are nested. If a model, H_0, is nested inside another, H_1, then H_0 is a subset of H_1, formed by some restriction imposed on H_1 (such as setting one or more parameters to zero). Nested model comparison is performed using the likelihood ratio test which states that under the null hypothesis (H_0) being correct, the likelihood ratio, $-2\ln[L(H_0)/L(H_1)]$, is distributed as $\chi^2(\nu)$, where ν is the number of additional parameters in H_1. If the models are not nested then the likelihood ratio cannot be used and we must resort to other methods. The Akaike information criterion (AIC) and the Bayesian information criterion (BIC) both penalize the likelihood by a term dependent on the number of parameters (Burnham and Anderson, 1998). Once this correction has been made the resultant values for each model can be compared and used to select the most appropriate to explain the data. The parametric bootstrap is another alternative and examines the empirical distribution of the likelihood ratio statistic by resampling (bootstrapping) methods (Efron and Tibshirani, 1998). The null model is fit to the data and then hypothetical data sets drawn from the fitted model. Both H_0 and H_1 models are fit to these samples and the empirical likelihood ratio distribution generated. If the observed likelihood ratio is greater than some quantile of the empirical distribution (say 95%) then H_1 is preferred over H_0.

Bayesian statistics

The Bayesian framework requires the specification of two objects; the likelihood and the prior (Gelman *et al.*, 2003). The likelihood we have already met (though it is now interpreted as a sampling probability distribution) and the prior distribution encodes our knowledge of the system before looking at the data. This can take the form of a simple range of values, in which case the prior is a uniform distribution, or a more complex distribution. The product of the prior and the likelihood (and suitably normalized) gives the posterior distribution. This object contains all the information about the inference and strikes a balance between what we can learn from the data, and what we knew or, and this has been the cause of statistical controversy, expected the system behaviour to be before we looked at the data. For example, this may include innocuous assumptions that chemical reaction rates have to be positive real numbers; on the other hand the prior may also include specific assumptions, for example about experimental error rates. Bayesian inference can be seen as the process of moving from the prior distribution to the posterior distribution in light of the data (which is encoded in the likelihood). There are many aspects of Bayesian inference that makes its use appealing in systems biology. The posterior, because it is a distribution, provides credible limits on the inferred parameter values and their correlation structure. For example, in our height data example, we no longer obtain just the optimal mean and variance (and possibly the associated confidence limits) but the probability of it taking that value. That is we can easily answer questions of the form, what is the range of values such that we are 95% certain that the mean is contained? Or, how correlated are my estimates of the mean and variance? As already mentioned these questions can be addressed in a non-Bayesian framework but they usually require further approximation or computationally demanding algorithms. Fully Bayesian methods, although generally more computationally intensive than traditional methods, provide this information for free.

The Bayesian framework also handles model selection in a very natural way that is simply an extension of parameter inference. The mathematical object of interest for model selection, the marginal likelihood, is the integral of the posterior distribution over the parameters. Thus the equivalent of the hypothesis test, the ratio of marginal likelihoods, known as the Bayes factor, automatically accounts for the difference in the number of parameters (since the dependence on the parameter values has been integrated out). This approach to model selection gives a posterior distribution over model space that is mathematically consistent and therefore easy to interpret. An additional benefit of obtaining probabilities from model selection is that we can make a weighted average of models in a technique known as Bayesian model averaging. This can be shown to always improve predictive power (though perhaps at a cost of losing explanatory power). Bayesian statistics formalizes our assumptions before inference through the specification of the prior distributions. When we perform parameter inference and the data are highly informative for the parameters of interest the effect of the prior assumptions become negligible, as the likelihood becomes more important than the prior. In contrast if the data are non-informative for a set of parameters their posterior distributions resemble the prior because we have not added any information to our knowledge. When performing model selection we should always interpret the results in light of our assumptions and examine how our conclusions change under different scenarios. While this 'subjective' nature of Bayesian statistics is occasionally criticized, the rigorous statement of assumptions should be seen as an advantage of Bayesian methods over traditional ones. The extra work we have to perform is inconsequential when compared with the huge benefits of the Bayesian approach.

As already discussed, inference in systems biology generally involves fitting high-dimensional models with many unknown parameters to low dimensional data with large experimental errors (Golightly and Wilkinson, 2006; Vyshemirsky and Girolami, 2008; Xu *et al.*, 2010). It is clear that the data can only be informative for a small subset of parameters (Gutenkunst *et al.*, 2007; Daniels *et al.*, 2008; Erguler and Stumpf, 2011). The subset of parameters for which information is available is generally not obvious from the model and the data *a priori*. Bayesian methods handle these situations in a consistent manner. Stochastic models pose a more challenging problem since it is rarely possible to write down the exact likelihood. However, in most of these cases it is possible to simulate the process given the parameters in the model. A family of techniques have been developed, known collectively as approximate Bayesian computation (ABC), that can perform simulation based inference without knowledge of the likelihood (Beaumont *et al.*, 2002; Toni *et al.*, 2009; Toni and Stumpf, 2010). These methods further extend the applicability of Bayesian inference to challenging inference problems in systems biology.

As an illustration we show in Fig. 1.3 the posterior distribution obtained from mRNA data in the repressilator model (see Fig. 1.3; the example is taken from the ABCSys-Bio package; Liepe *et al.*, 2010; Zhou *et al.*, 2011). Not all parameters can be inferred from this type of data. This phenomenon has now been established comprehensively: for any given experimental set-up only a small number of parameters can be inferred (or estimated) with any degree of certainty. But while the bulk of parameters cannot be estimated it can also be shown, that those parameters that have a large influence on a system's dynamics are also the ones that can be inferred. Thus parameters that cannot be inferred generally do not matter much (at least under the present conditions). It is, however, also possible to design experimental set-ups that allow for the inference of other parameters.

Comparative approaches

All living organisms share common properties and are all related to each other through a common shared phylogeny. Evolutionary processes such as mutation, recombination, drift

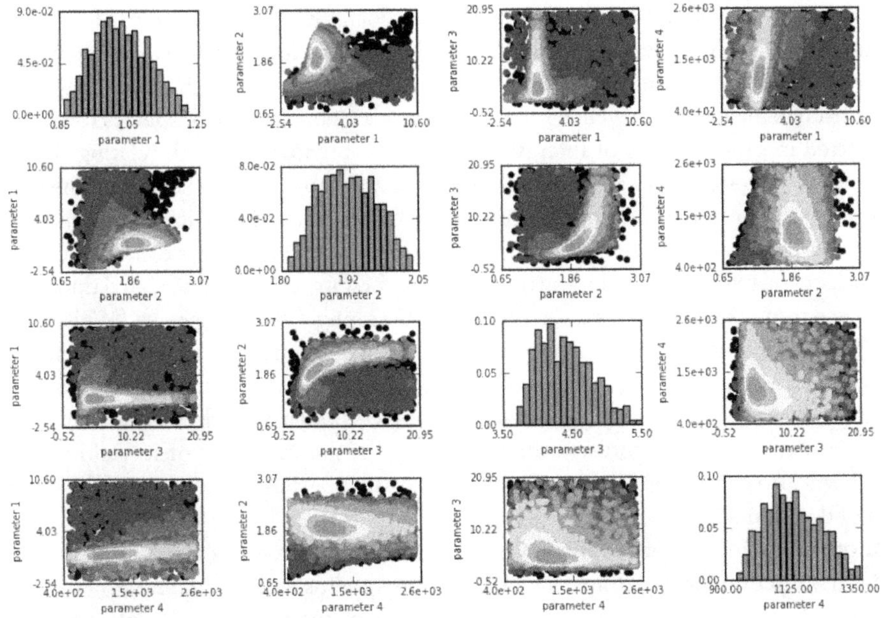

Figure 1.3 Posterior probability surfaces of the deterministic repressilator model (see Toni *et al., 2009*). The scatter plots show pairs of parameters plotted against one another, whereas the histograms reflect the probability densities. Shown are the best estimates for the four parameters and the intermediate distributions which are obtained as part of the sequential inference procedure (Liepe *et al.*, 2010). In the scatterplots the extent of the dots along the horizontal and vertical axis ranges give an idea of how well the parameters are identified, and whether they are statistically dependent. Parameter 4 is only poorly inferred, while parameter 2, a Hill coefficient of the model, is very well localized in the ABC procedure.

and selection operate at the population level (Lynch, 2007). As subpopulations from within a species diverge from one another, they retain high levels of genetic similarity in their coding and, though to a generally lesser degree, their non-coding sequence. But eventually these different subpopulations can form separate species, by which we typically mean that individuals from different species are unable to produce viable offspring. These different species still share large parts of their DNA and protein sequence; for bacteria (and viruses) the species concept is, of course, somewhat less well defined, but as subpopulations form recognizable strains, and eventually different bacterial species, their genomic sequence still allows us to trace and indeed reconstruct their evolutionary ancestry.

Generally, all organisms tend to share common properties with, usually, small variations in between closely related species. As an example, all living organisms share the process through which DNA is replicated. This link relating all extant life provides a valuable source of information for developing models of biological systems (Stumpf *et al.*, 2007). In the following paragraphs three different ways to use this information are presented relating to different level of simplification/reduction of the modelling space by (i) defining an initial starting point for a model, (ii) reducing the space to explore, or finally (iii) allowing us to transfer experimental data obtained from one species/organisms to make statements about processes in another species.

In the last decades a lot of effort has been invested in the analysis of few model organisms including Escherichia coli, Saccharomyces cerevisiae, etc. However, these model organisms are not the only organisms of interest and even these species, though extensively studied, are far from being completely characterized. Discoveries made in these species can be used/transferred in other species of interest based on the phylogenetic link relating them. If a given feature is observed in different species and a mechanistic model exists in one species, then this can typically serve as a starting point for mechanistic, or indeed mathematical, analysis of the same process in the species of interest. In bacteria, for example, metabolic networks, gene regulation, including transcription-factor specificity, and signal transduction and stress response mechanisms are often shared among different species (Wuichet *et al.*, 2010). As a guiding rule, evolutionary separation is a good measure as to how similar such systems are. But generally the relationship is more subtle than for example protein sequence similarity, where this relationship between sequence similarity and evolutionary distance is generally relatively straightforward. The analysis of the variations observed in different species will also give valuable insight into the system properties, often allowing for improvement of the model.

Another possibility opened up through comparative approaches is the reduction of the model and its complexity: comparing orthologous systems in different species (i.e. systems that consist of gene products that have diverged following speciation events) may enable us to identify cores of such systems that facilitate the functionality for example of a stress response or metabolic pathway (Huvet *et al.*, 2011). These in turn may be easier to model or describable in terms of simpler models than the potentially more refined or nuanced extant mechanisms observed in model species.

Finally, the ancestral relationships among species allows, to a certain extent, the aggregation of experimental data obtained in different species for a given system. Sometimes the amount of experimental data available for a given system in a given species is just not sufficient to allow modelling to be performed (too many parameters for the available data for example). If we assume that a system is well conserved between different species and is derived from the same ancestral system, then, again with care and caution, we may pool insights gained in different systems to build a single integrated model. This type of approach will lead to the creation of an 'averaged' system but could provide a starting point helping for identifying discriminatory experiments. We might expect this approach to give rise to qualitative rather than quantitatively predictive models. Nevertheless this may serve as a starting point for more detailed investigations.

A few caveats are in order, however: orthology at the sequence level is easier to establish than at the system level; furthermore, we generally lack reliable mappings from genotype (DNA sequence) to phenotype (such as stress response or metabolic phenotypes). The use of the evolution relationship between biological systems and organisms seems to be an easy way to improve modelling accuracy. However, the use of this type of information is not quite as straightforward because of the lack of knowledge about the links between the genotype (DNA sequences) and the phenotype (observable properties of an organism). While we have to be cautious when applying comparative approaches in the development of mathematical models, the information is probably too useful and valuable to be discarded. Quite generally, we believe that evolutionary approaches have a lot to offer to systems and synthetic biology.

Text-mining

Text-mining is the final technique that can be used in order to construct mathematical models of biological systems. The scientific literature contains a myriad of information which is useful in the development of models of prokaryotic processes and systems. However, as the publication rate continues to grow (exponentially or perhaps even higher) it is impossible for researchers to maintain an adequate awareness of the literature that is relevant for them. This has led to the development of text-mining systems for use in the biological and biomedical domain. Text-mining combines techniques from a variety of fields including information retrieval (IR), machine learning (ML), natural language processing (NLP), statistics and computational linguistics (CL) in order to reduce the cost and time of extracting information from text. Text mining can be thought of as a method to perform a systematic review over large numbers of papers. It has several advantages over manually performing a review: speed, cost and is scalable.

The task of identifying entities of interest (such as molecules) in text is called named entity recognition (NER). These entities of interest can be genes, proteins, chemicals, species names, cellular components etc.. However the biological domain presents many and very particular challenges for developers of NER systems to deal with. The nomenclature is highly variable and therefore simple NER systems where potential named entities are matched against a pre-defined dictionary tend to perform poorly. More complicated ml approaches such as support vector machines (SVMs) and conditional random fields (CRFs) have been found to attain good performance in identifying gene and protein names in text.

Once the named entities (NEs) have been identified they need to be normalized to canonical identifiers (e.g. gene names need to be mapped to Entrez gene identifiers). However this is not a simple one to one mapping problem and one gene name can refer to many Entrez gene identifiers in one or many species. Several approaches have been proposed to distinguish between multiple identifiers from using simple rules to generate variations of NEs to more complicated ml approaches.

Within the more general systems biology and bioinformatics domains text mining is being used in an assortment of ways (Harmston *et al.*, 2010). The content of several resources as been augmented with information gained from text mining [such as the STRING database (von Mering *et al.*, 2005) and its use has led to increases in performance in predicting subcellular locations of proteins (Stapley *et al.*, 2002; Lu *et al.*, 2004), network inference (Li *et al.*, 2010) and identifying genes involved in disease (Al-Mubaid and Singh, 2010). Some recent work in text mining has focused on identifying terms involved in the identification of named entities (Ananiadou *et al.*, 2011) related to Type IV secretion systems (T4SS).

An important part of systems biology is the generation of mathematical models of biological systems, the investigation of their properties, the generation of predictions and the experimental validation or rejection of such proposed models in order to gain understanding about the system. A deterministic model of a biochemical system can be generated from two components: a stoichiometry matrix, S, and a rate vector, v. The stoichiometry matrix (S) holds information on the molar proportions of reactants and products for individual reactions in a biochemical system X. It is a m row and n column matrix where m is the number of reactants in the entire system and n is the number of individual reactions. The rate vector (v) contains the reaction rates of the n reactions.

Given these two components it is then possible to generate differential equations for each reaction using the following equation (assuming mass-action kinetics):

$$\frac{dX}{dt} = Sv \tag{1.6}$$

The generation of the stoichiometric matrix, through the identification of the reactants and reactions that are involved in a biological system of interest and the determination of individual reaction rates requires considerable investment in terms of time and effort. In order to build a good model, only a small proportion of the available literature needs to be investigated, however identifying this set is difficult in the first place. A biologist would have to identify these papers, read and extract from them the necessary information to build a good model.

Given a set of genes that we are interested in modelling it is possible to use text mining to identify relevant papers containing information on these genes as well as proposing further genes which may also be involved in the pathway of interest. By looking at the co-occurrence of gene mentions in documents and sentences it is possible to generate a list of candidate genes, and if they are relevant use these to expand the constituent parts of the model. More complicated methods are currently being investigated where the relationship between two co-occurring entities can be determined and used to propose entries in the stoichiometry matrix directly. Of course, eventually all the papers still have to be read to check for relevance, but the identification step of the model generation process is greatly sped up.

Some success has been had with the development of text-mining tools to identify papers containing information of parameters (Hakenberg *et al.*, 2004) and in identifying individual parameter values. The KiD database was generated using text mining to extract parameters of biochemical systems from the abstracts in Medline (Heinen *et al.*, 2010). The automated extraction of such parameters enables them to be included in of these in the rate vector directly or used in weakly informative priors for Bayesian inference of parameters.

Text mining, like comparative approaches, offers a way for to decrease the development time of new models, allowing experimental results to be collated in an easily searchable way. The ability to help inform the contents of the model and the parameters associated with that model helps improve the efficiency of researchers involved in model generation.

Conclusion

Making models requires crucially not just competence with the reverse engineering tools and methodologies, but detailed knowledge of the biological systems under consideration. This in turn requires adapting modelling approaches to the biological problem. In the past mathematical biologists have sometimes focused on a given mathematical approach, such as ordinary or partial differential equations, and applied it to a suitable set of biological problems. Increasingly, however, a shift of focus can be observed, where the biological problem at hand dictates the use of certain mathematical techniques. Theoretical modellers thus have to become generalists and be conversant with different modelling frameworks and with the statistical tools required to calibrate models against data. Frequently changes in experimental technology will pose new challenges to statistics, which can for example be witnessed as single cell data (or single molecule data) are becoming more prevalent. When this happens it frequently also means that past publications become less informative.

Modelling is relatively straightforward compared with model development. While the level of mathematical sophistication in the life- and biomedical sciences continues to increase, it is the dialogue between modellers and biological domain experts that has really been advancing the field. At the moment we lack reliable models, or mechanistic insights, for most biological processes of interest or importance. The three complementary frameworks outlined above, (i) statistical analysis of data, (ii) comparative and evolutionary studies, and (iii) text mining to capture the wealth of available information in the literature, can be used to come up with new or improved models. The precise weighting that should be given to each approach will depend on the situation at hand.

References

Al-Mubaid, H., and Singh, R. (2010). A text-mining technique for extracting gene-disease associations from the biomedical literature. Int. J. Bioinform. Res. Appl. *6*, 270–86.

Alm, E., and Arkin, A.P. (2003). Biological networks. Curr. Opin. Struct. Biol. *13*, 193–202.

Ananiadou, S., Sullivan, D., Black, W., Levow, G.A., Gillespie, J.J., Mao, C., Pyysalo, S., Kolluru, B., Tsujii, J., and Sobral, B. (2011). Named entity recognition for bacterial type iv secretion systems. PLoS ONE *6*, e14780.

Alon, U. (2007). Network motifs: theory and experimental approaches. Nat. Genet. Rev. *8*, 450–461.

Beaumont, M., Zhang, W., and Balding, D. (2002). Approximate Bayesian computation in population genetics. Genetics *162*, 2025–2035.

Bornholdt, S. (2005). Systems biology: less is more in modeling large genetic networks. Science *310*, 449.

Bornholdt, S. (2008). Boolean network models of cellular regulation: prospects and limitations. J. Roy. Soc. Interface *5*, S85—94.

Burnham, K., and Anderson, D. (1998). Model Selection and Multimodel Inference (Springer, New York).

Cox, D. (2006). Principles of Statistical Inference (Cambridge University Press, Cambridge).

Cox, D., and Hinkley, D. (1974). Theoretical Statistics (Chapman & Hall/CRC, New York).

Daniels, B.C., Chen, Y.J., Sethna, J.P., Gutenkunst, R.N., and Myers, C.R. (2008) Sloppiness, robustness, and evolvability in systems biology. Curr. Opin. Biotech. *19*, 389–395.

Efron, B., and Tibshirani, R. (1998) An Introduction to the Bootstrap (Chapman & Hall, New York).

Elowitz, M.B., and Leibler, S. (2000). A synthetic oscillatory network of transcriptional regulators. Nature *403*, 335–8.

Elowitz, M.B., Levine, A.J., Siggia, E.D., and Swain, P.S. (2002). Stochastic gene expression in a single cell. Science *297*, 1183–1186.

Erguler, K., and Stumpf, M.P.H. (2008). Statistical interpretation of the interplay between noise and chaos in the stochastic logistic map. Math. Biosci. *216*, 90–99.

Erguler, K., and Stumpf, M.P.H. (2011). Practical limits for reverse engineering of dynamical systems: a statistical analysis of sensitivity and parameter inferability in systems biology models. Mol. Biosyst. *7*, 1593–1602.

Gelman, A., Carlin, J.B., Stern, H., and Rubin, D. (2003) Bayesian Data Analysis, 2nd edn (Chapman & Hall/CRC, Boca Raton).

Golightly, A., and Wilkinson, D.J. (2006). Bayesian sequential inference for stochastic kinetic biochemical network models. J. Comput. Biol. *13*, 838–851.

Gunawardena, J. (2010). Biological systems theory. Science *328*, 581.

Gutenkunst, R.N., Waterfall, J.J., Casey, F.P., Brown, K.S., Myers, C.R., and Sethna, J.P. (2007). Universally sloppy parameter sensitivities in systems biology models. PLoS Comp. Biol. *3*, 1871–1878.

Hakenberg, J., Schmeier, S., Kowald, A., and Klipp, E. (2004). Finding kinetic parameters using text mining. Omics *8*, 131–152.

Harmston, N., Filsell, W., and Stumpf, M.P.H. (2010). What the papers say: text mining for genomics and systems biology. Hum. Genom. *5*, 17–29.

Hastie, T., Tibshirani, R., and Friedman, J. (2003). The Elements of Statistical Learning (Springer, New York).

Heinen, S., Thielen, B., and Schomburg, D. (2010). Kid – an algorithm for fast and efficient text mining used to automatically generate a database containing kinetic information of enzymes. BMC Bioinfo. *11*, 375.

Huvet, M., Toni, T., Sheng, X., Thorne, T., Jovanovic, G., Engl, C., Buck, M., Pinney, J.W., and Stumpf, M.P.H. (2011). The evolution of the phage shock protein response system: interplay between protein function, genomic organization, and system function. Mol. Biol. Evol. *28*, 1141–1155.

Ingram, P.J., Stumpf, M.P.H., and Stark, J. (2008). Nonidentifiability of the source of intrinsic noise in gene expression from single-burst data. PLoS Comp. Biol. *4*, e1000192.
Kelly, W.P., and Stumpf, M. (2008). Protein–protein interactions: from global to local analyses. Curr. Opin. Biotechnol. *19*, 396–403.
Kholodenko, B.N., Hancock, J.F., and Kolch, W. (2010). Signalling ballet in space and time. Nat. Rev. Mol. Cell Biol. *11*, 414.
Kussel, E., and Leibler, S. (2005). Phenotypic diversity, population growth, and information in fluctuating environments. Science *309*, 2075–2078.
Lèbre, S., Becq, J., Devaux, F., Stumpf, M.P.H., and Lelandais, G. (2010). Statistical inference of the time-varying structure of gene-regulation networks. BMC Syst. Biol. *4*, 130.
Li, Y., Hu, X., Lin, H., and Yang, Z. (2010). Learning an enriched representation from unlabeled data for protein–protein interaction extraction. BMC Bioinfo. *11*, S7.
Liepe, J., Barnes, C., Cule, E., Erguler, K., Kirk, P., Toni, T., and Stumpf, M.P.H. (2010). ABC-SysBio–approximate Bayesian computation in Python with GPU support. Bioinformatics *26*, 1797–1799.
Lu, Z., Szafron, D., Greiner, R., Lu, P., Wishart, D.S., Poulin, B., Anvik, J., Macdonell, C., and Eisner, R. (2004). Predicting subcellular localization of proteins using machine-learned classifiers. Bioinformatics *20*, 547–556.
Lynch, M. (2007). The evolution of genetic networks by non-adaptive processes. Nat. Rev. Genet. *8*, 803–813.
May, R.M. (2004). Uses and abuses of mathematics in biology. Science *303*, 790–793.
von Mering, C., Jensen, L., Snel, B., and Hooper, S. (2005). String: known and predicted protein–protein associations, integrated and transferred across organisms. Nucl. Acids Res. 33, D433–477 .
Nurse, P., and Hayles, J. (2011). The cell in an era of systems biology. Cell *144*, 850–854.
de Silva, E., and Stumpf, M. (2005). Complex networks and simple models in biology. J. Roy. Soc. Interface *2*, 419–340.
Stapley, B., Kelley, L., and Sternberg, M. (2002). Predicting the sub-cellular location of proteins from text using support vector machines. Pac. Symp. Biocomput. 374–385.
Stigler, B., Jarrah, A., Stillman, M., and Laubenbacher, R. (2007). Reverse engineering of dynamic networks. Ann. N. Y. Acad. Sci. 1115, 168–177.
Stricker, J., Cookson, S., Bennett, M.R., Mather, W.H., Tsimring, L.S., and Hasty, J. (2008). A fast, robust and tunable synthetic gene oscillator. Nature *456*, 516–519.
Stumpf, M.P.H., Kelly, W.P., Thorne, T., and Wiuf, C. (2007). Evolution at the system level: the natural history of protein interaction networks. Trends Ecol. Evol. *22*, 366–373.
Toni, T., and Stumpf, M.P.H. (2011). Simulation-based model selection for dynamical systems in systems and population biology. Bioinformatics *26*, 104–110.
Toni, T., Welch, D., Strelkowa, N., Ipsen, A., and Stumpf, M.P. (2009). Approximate Bayesian computation scheme for parameter inference and model selection in dynamical systems. J. Roy. Soc. Interface *6*, 187–202.
Toni, T., Jovanovic, G., Huvet, M., Buck, M., and Stumpf, M.P. (2011). From qualitative data to quantitative models: analysis of the phage shock protein response in *Escherichia coli*. BMC Syst. Biol. *5*, 69.
Vyshemirsky, V., and Girolami, M.A. (2008). Bayesian ranking of biochemical system models. Bioinformatics *24*, 833–839.
Wilkinson, D. (2006) Stochastic Modelling for Systems Biology (Chapman & Hall/CRC, Boca Raton).
Wilkinson, D. (2009) Stochastic modelling for quantitative description of heterogeneous biological systems. Nat. Rev. Genet. *10*, 122–133.
Wuichet, K., Cantwell, B.J., and Zhulin, I.B. (2010). Evolution and phyletic distribution of two-component signal transduction systems. Curr. Opin. Microbiol. *13*, 219–225.
Xu, T.R., Vyshemirsky, V., Gormand, A., von Kriegsheim, A., Girolami, M., Baillie, G.S., Ketley, D., Dunlop, A.J., Milligan, G., Houslay, M.D., *et al.* (2010). Inferring signaling pathway topologies from multiple perturbation measurements of specific biochemical species. Science Signal. *3*, ra20.
Zhou, Y., Liepe, J., Sheng, X., Stumpf, M.P.H., and Barnes, C. (2011). GPU accelerated biochemical network simulation. Bioinformatics *27*, 874–876.

Dynamics and Robustness of Metabolic Networks: A Systems Biology Review of *Escherichia coli* Metabolism

2

Eivind Almaas, Per Bruheim, Rahmi Lale and Svein Valla

Abstract

The functional repertoire of an organism's metabolic network is closely linked to its phenotype and potential for utility in metabolic engineering applications. In this chapter, we discuss a systems biology view of *Escherichia coli* metabolism by integrating current genome-scale computational modelling approaches with available molecular genetics tools, as well as the experimental framework for metabolite and metabolic flux determination.

Introduction

The advent of high-throughput technologies has changed microbiology from a primarily 'reductionist' field of science, with a focus on one specific cellular process or pathway, to one in which the biology of the system as a whole is studied. To fully achieve this, detailed biological data from all cellular levels (i.e. genomic, transcriptomic, proteomic, and metabolomic) should be utilized. However, this is a significant challenge since analytical techniques for sequencing and measuring the expression levels of complete genomes are far more advanced than similar technologies for analyses of a cell's protein and metabolite complements. The revolutionary progress in genome sequencing and other high-throughput methods has given rise to genome-level models of metabolism that have been analysed using constraint-based approaches. The predictions from these computational tools can be directly related to methods for molecular genetics manipulation of cellular metabolism and experimental determination of metabolic fluxes.

Computational modelling of *E. coli* metabolism

Currently, the family of constraint-based approaches is the most widely used computational method for studying genome-scale metabolism. Among the constraint-based methods, flux-balance analysis (FBA) (Edwards and Palsson, 2000a; Palsson, 2006) is most frequently applied. The success of constraint-based approaches stems from the fact that, unlike kinetic models, they aim to identify optimal metabolic steady-state activity patterns that satisfy constraints imposed by mass balance, the metabolic network structure, and the availability of nutrients (Palsson, 2006). In other words, the imposed constraints limit the flux ranges individual reactions may access and do not identify a particular flux solution for the system. Optimization is invoked to identify flux states that may match experimental conditions and designs (Palsson, 2006).

However, FBA is only one among many constraint-based approaches. Other important examples include minimization of metabolic adjustment (MOMA) (Segre *et al.*, 2002), Regulatory on/off minimization (ROOM) (Shlomi *et al.*, 2005), and the possibility to include gene- and protein-expression data together with the metabolic network optimization (Shlomi *et al.*, 2008). The growth of computational approaches for the predictive modelling of cellular metabolic phenotypes and performance, combined with the rapidly increasing availability of high-quality metabolic reconstructions (Henry *et al.*, 2010), has transformed the *in silico* analysis of cellular metabolism to an invaluable part of systems biology.

Principles for flux-balance analysis and experimental support

In this section we will go through the principles of FBA in detail, and describe its strengths, shortcomings, as well as experimental support for this approach.

Principles for flux-balance analysis

Flux-balance analysis (FBA) is a modelling approach that is based on the genome-scale metabolic reconstruction of an organism, encoded by the stoichiometric matrix, S $(m \times n)$, where m is the number of metabolites and n the number of reactions (Palsson, 2006; Orth *et al.*, 2010). When applying the reasonable assumption of mass balance to every metabolite present, we need to tally up every instance of a metabolite M_is production and consumption:

$$\frac{dM_i}{dt} = \sum_{j=1}^{n} S_{ij} v_j \qquad (2.1)$$

Here, v_j corresponds to the metabolic flux through reaction j. The fluxes depend in general on the concentrations in a non-linear fashion. One challenge in specifying these non-linear relationships between the reaction fluxes and metabolite concentrations is the necessity to introduce reaction rate coefficients. While some may be determined *in vitro*, it is not obvious that these results are reasonable for the *in vivo* conditions.

The FBA approach completely sidesteps this issue by making the steady-state approximation, where all time derivatives are zero. As a result, we are left with the following system of linear equations in the metabolic fluxes v_j:

$$\sum_{j=1}^{n} S_{ij} v_j = 0 \qquad (2.2)$$

Since S is not a square matrix, there exists a multitude of solutions that satisfy eqn 2.2. We may reduce the degeneracy in the possible values that flux solutions to eqn 2.2 may take by adding limitations on their ranges based on experimental studies. For instance, we may determine the maximal and minimal metabolic flux through a reaction, the flux of nutrients imported, or waste products from an organism:

$$\alpha \leq v_i \leq \beta, \qquad (2.3)$$

$$\chi \leq b_i \leq \phi \tag{2.4}$$

where v_i and b_i are the internal export/import flux of species *i* respectively, and α, β, χ and ϕ are the lower and upper limits for these fluxes.

FBA typically combines the above framework with linear programming in order to identify metabolic flux states of biological relevance. While cellular biomass is typically chosen as the objective function to maximize, there is a growing body of evidence that an appropriate objective function depends on the growth conditions of the experiment (e.g. the presence of nutrient limitations) (Feist and Palsson, 2010). Other possibilities for the objective function include, but are not limited to, maximization of biomass yield (Schuetz *et al.*, 2007), minimization of redox potential (Knorr *et al.*, 2007; Schuetz *et al.*, 2007), maximization of ATP yield per flux unit (Schuetz *et al.*, 2007), and maximization of ATP consumption (Ow *et al.*, 2009). For a recent review of studies of different objective functions (see Feist and Palsson, 2010).

Experimental support for FBA

The first genome-scale metabolic model for *E. coli*, iJE660 (Edwards and Palsson, 2000a), systematically investigated the computationally predicted single-gene knockout phenotypes for *E. coli* growth in four minimal environments, either with glucose, glycerol, acetate or succinate as the sole carbon source. This study demonstrated that 86% of the computationally predicted single-gene knockout growth phenotypes were in agreement with experiments (Edwards and Palsson, 2000a). Later studies of a high-quality genome-scale model of yeast (Duarte *et al.*, 2004) found a similar level of agreement (84%) between experimentally tested gene knockouts and computational predictions from FBA.

As discussed above, a central element in the constraint-based methodology is optimization of an objective function. Three landmark articles convincingly demonstrated that *E. coli* metabolic capabilities are well described by the FBA approach using optimization of biomass. In the first article (Edwards *et al.*, 2001), the authors described how they tested if *E. coli* in fact was utilizing the metabolic network optimally in order to achieve maximal growth by simultaneously measuring (i) cellular growth rate, and the uptake rate of (ii) a single carbon source and (iii) oxygen. For both primary carbon sources tested (acetate and succinate), the FBA predicted optimal ratios between growth rate, carbon source uptake rate and oxygen uptake rate were in good agreement with experimental observations for acetate (Edwards *et al.*, 2001): The overall average error between the observed and the predicted uptake fluxes (including growth rate) was found to be 5.8%. However for succinate, the overall average error was found to be 10.7%.

The larger level of error between FBA predicted metabolic phenotypes and growth experiments using succinate as the primary carbon source could be caused by incomplete adaptation by *E. coli* to the nutrient environment, i.e. that *E. coli* grows suboptimally. The hypothesis of suboptimal growth was investigated in the second article (Ibarra *et al.*, 2002), where *E. coli* was placed under growth-selective pressure in minimal media with acetate, succinate, malate, glucose or glycerol as the primary carbon source. Over a minimum of only 500 generations, the maximal growth rate was found to increase 17–20% for these carbon sources. Furthermore, when tracking trajectories of the growth rate, the oxygen and primary carbon uptake rates, they all moved towards FBA predicted optimal values (Ibarra *et al.*, 2002).

The third landmark article investigated if the previously found ability of FBA to accurately predict optimal metabolic performance also was valid for mutant strains of *E. coli* with in-frame deletions of metabolic genes (Fong and Palsson, 2004). The genes chosen for deletion were *ackA* (acetate kinase A), *frd* (fumarate reductase), *pck* (phosphoenolpyruvate carboxykinase), *ppc* (phosphoenolpyruvate carboxylase), *tpi* (triosephosphate isomerase) and *zwf* (glucose 6-phosphate-1-dehydrogenase), and each mutant strain was evolved in a minimal medium with four different primary carbon sources. The study found that mutant strains increased their growth rates by an average of 87% during adaptive growth. The final growth rates for the mutant strains were within 10% of the computationally predicted rates in 78% of the tested cases (Fong and Palsson, 2004). This result underscores the importance of and the need for an organism to be fully adapted to its growth environment in order for constraint-based approaches to be predictive; both wild-type organisms and mutant strains.

In order to assess the accuracy of constraint-based modelling in general, and FBA in particular, it is necessary to compare model predictions with experimentally determined metabolic fluxes. While much evidence exists for good agreement in FBA flux predictions in central carbon metabolism (Palsson, 2006), here we will mention in brief comparisons between experiments (Fischer *et al.*, 2004) and one genome-scale reconstruction, designated iAF1260 (Feist *et al.*, 2007). Typically, experimentally determined carbon and oxygen uptake rates are used as modelling constraints, leaving internal fluxes, growth rate, and secretion fluxes to be determined by FBA (Palsson, 2006). For *in silico* model iAF1260, FBA predicted growth rate, average glucose uptake rate, average oxygen uptake rate, and CO_2 production rates very well (Feist *et al.*, 2007). Additionally, comparisons with experimentally determined fluxes for central carbon metabolism (Fischer *et al.*, 2004) and FBA predictions showed complete agreement in flux direction through many pathways, including the TCA cycle, glycolytic pathway, and pentose phosphate pathway (Feist *et al.*, 2007). On a direct comparison of 22 experimentally determined central carbon metabolism flux values (Fischer *et al.*, 2004), it was reported $R^2 = 0.96$ (Fischer *et al.*, 2004).

Principles for genome-scale metabolic reconstruction

With the ever increasing availability of high-throughput 'omics' data, it has become a reality to develop a high-quality genome-scale model of an organism's metabolic network relatively quickly and with a high degree of automation (Henry *et al.*, 2010). Recently, several articles have discussed in great detail the different steps involved in generating a high-quality metabolic reconstruction and their complexity (Thiele and Palsson, 2010), as well as reviewed various aspects of the model building process (Reed and Palsson, 2003; Feist and Palsson, 2008; Feist *et al.*, 2009; Oberhardt *et al.*, 2009). Although many time-consuming steps in the model-building process may be implemented by automatic, unsupervised software programs, there are important aspects of the reconstruction process that still have to be manually curated (Henry *et al.*, 2010).

In short, the process of generating a high-quality metabolic reconstruction can be divided into four main steps (Feist *et al.*, 2009; Thiele and Palsson, 2010): (i) automated reconstruction based on an annotated genome, (ii) further curation and refinement of reconstruction, (iii) converting the reconstruction into a computational model, and (iv) computational analysis and integration of the reconstruction with high-throughput 'omics data. In the first step in the reconstruction process, the focus is to identify gene products that act as metabolic enzymes or take part in protein complexes with enzymatic function.

Using combinations of existing, well-maintained databases such as EntrezGene (Maglott *et al.*, 2007), Genome Reviews (Kulikova *et al.*, 2004), KEGG (Kanehisa and Goto, 2000), and MetaCyc (Krieger *et al.*, 2004), it is possible to automatically convert a collection of annotated genes to a listing of reactions catalysed by enzymes (Henry *et al.*, 2010).

While the first step is well suited to automated data extraction and reconstruction approaches, the second step of refinement is currently more of a challenge (Thiele and Palsson, 2010). Important components of step (ii) are to add organism-specific reactions, including determination of reaction direction and stoichiometry, subsystem localization, filling of gaps in pathways, and the determination of biomass composition. However, recent progress in algorithms for automated model reconstruction has also made many of the components of step (ii) amenable (Henry *et al.*, 2010). As we will discuss in the Prediction of gene knockout section, an accurate determination of the components of the biomass reaction is directly related to the computational prediction of gene-knockout phenotypes.

The third step in the model reconstruction process is concerned with generating a representation that is accessible to computational analysis. As of the present time, there exists several software packages for the *in silico* analysis and experimentation on metabolic reconstructions (Becker *et al.*, 2007; Rocha *et al.*, 2010; Hoppe *et al.*, 2011) which all accept metabolic reconstructions represented in the systems biology mark-up language (SBML) (Hucka *et al.*, 2003). The software packages contain different functionalities (Hoppe *et al.*, 2011) and approaches to user interface. With the use of one or more program suites and an SBML encoded model, it is thus possible to enter step (iv) of the metabolic reconstruction process. Here, the computational predictions of the model's ability to produce biomass precursors in a wide range of environments, growth rates and growth yields, and gene-deletion phenotypes can be tested against experimental data. Through an iterative process, identified discrepancies between computational predictions and experimental data can be used to improve the model properties and add new components to a model in step (ii) of the reconstruction process (Covert *et al.*, 2004; Thiele and Palsson, 2010).

Prediction of gene knockout phenotypes using constraint-based modelling

The access to a genome-scale metabolic model that includes gene-protein-reaction (GPR) relations makes it relatively straightforward to use constraint-based approaches to predict gene knockout phenotypes. After a gene or a set of genes has been selected for *in silico* knockout, one uses the GPR relations to identify which (if any) reactions will be affected. An example of a GPR relation is that of phosphoglucomutase (EC 5.4.2.2): In metabolic model iJR904 (Reed *et al.*, 2003), this is given as

$$\text{PGMT} - (\text{b0688 or b2690}) \tag{2.5}$$

indicating that the reaction PGMT can be catalysed by the enzyme resulting from either gene b0688 or b2690; they are isozymes. Thus, by only knocking out either gene b0688 or b2690 in the computer model reaction PGMT will be unaffected, and it will be necessary to remove both genes from the model to have an effect on reaction PGMT.

When using any constraint-based method, the computer model of a mutant strain is implemented by restricting the metabolic flux to zero in all reactions affected by the single or multiple gene knockouts. This mutant strain model may then be used as input for the computational phenotype determination (Palsson, 2006). To date, three approaches have received much attention for the computational simulation of gene knockouts: FBA (Palsson, 2006), minimization of metabolic adjustment (MOMA) (Segre *et al.*, 2002), and regulatory on/off minimization (ROOM) (Shlomi *et al.*, 2005). Fig. 2.1 shows a simplified representation of how these methods work. At the current time, the constraint-based software packages OptFlux (Rocha *et al.*, 2010) and FASIMU (Hoppe *et al.*, 2011) have encoded the FBA, MOMA and ROOM approaches for gene knockout simulations, while the COBRA Toolbox (Becker *et al.*, 2007) includes FBA and a linear version of MOMA among the functional repertoire.

FBA prediction of mutant phenotypes

The maybe simplest approach to investigate *in silico* gene knockout phenotypes is the direct application of FBA to the mutant strain model (Edwards and Palsson, 2000b; Palsson, 2006). However, using FBA directly implies that the mutant strain is expected to be fully adapted to a growth environment. As has been discussed above (Fong and Palsson, 2004), it is necessary with a period of competitive growth of the mutant strains for the FBA method to be applicable. The typical objective function is biomass production. Hence, an FBA gene knockout simulation is measuring the ability of the mutant metabolic network reconstruction to produce all the biomass components in a given environment. We may classify genes as always essential or conditionally essential, if their phenotypic effect depends on the nutrient

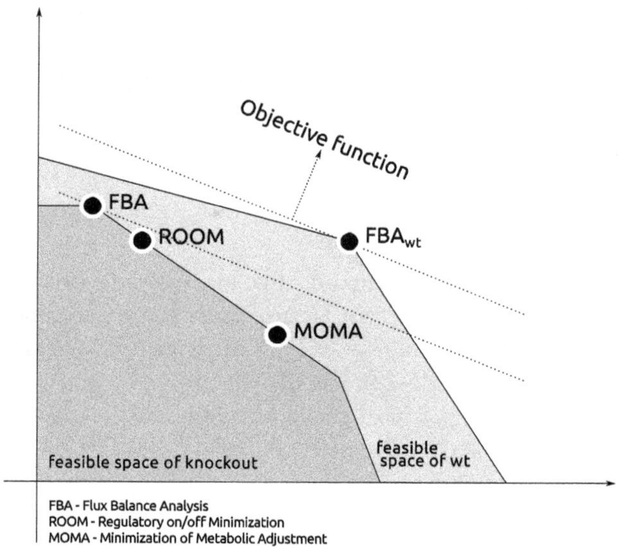

Figure 2.1 Drawing of constraint-based optimization principles. The light grey shaded region indicates the feasible flux space for the unperturbed (wild type) organism metabolic network. The dark shaded region indicates the part of the wild-type feasible space still available (feasible) after gene knockouts. FBA$_{wt}$ denotes the point in the feasible flux space where the wild type achieves optimal metabolic performance given the indicated objective function, while the point marked FBA denotes the maximally performing mutant. The point MOMA marks the closest location in the mutant feasible space to that of the FBA$_{wt}$ point. ROOM suggests a location where the fewest reaction fluxes have been changed from their wild-type values.

environment (Almaas *et al.*, 2005): always essential genes result in mutant reconstructions without the ability to grow when all available transport and substrate uptake reactions are activated, while conditionally essential genes result in non-growing mutants only in certain conditions.

The procedure to investigate the knockout phenotype in metabolic reconstruction strains with multiple simultaneous gene knockouts is similar: for each gene to be removed from the metabolic reconstruction, one needs to identify the affected reactions and constrain their fluxes to zero (Palsson, 2006). An FBA analysis with growth as the objective function will determine the ability of the multigene knockout mutant's metabolic network to produce biomass components in a given nutrient environment. This can result in quite surprising insights regarding the robustness and organization of central metabolism (Nakahigashi *et al.*, 2009).

MOMA prediction of mutant phenotypes

Minimization of metabolic adjustment (MOMA) was proposed as an alternative approach to FBA for the *in silico* prediction of mutant strain growth phenotypes (Segre *et al.*, 2002). While FBA intends to predict the growth phenotype of knockout mutants after adaptation to the nutrient environment has been achieved, MOMA intends to predict the growth phenotype before adaptation to the gene loss has occurred. Note, however, that MOMA presumes that the wild-type cell has fully adapted to the environment before gene knockouts are implemented (Segre *et al.*, 2002). The idea of MOMA is very simple: in the space of all reaction fluxes, the optimally performing wild type as determined by FBA is represented as a single point. After one or several genes have been removed from the metabolic network, the size and shape of the accessible flux space has been altered, and MOMA identifies the

Table 2.1 Tabulation of existing *E. coli* genome-scale metabolic network characteristics

Model	iJE660[1]	iJR904[2]	iAF1260[3]	iCA1273[4]
RFs	660	904	1260	1273
Metabolites	436	625	1039	1111
Reactions	720	931	2077	2477
Transport reactions[5]	140	143	304	317
Cellular compartment[6]	2	2	3	3
Computed growth rate[7]	0.915	0.922	0.929	0.905
Number of lethal genes[8]	160	174	187	191
Number of always lethal genes[9]	68	84	72	74

[1]Edwards and Palsson (2000a).
[2]Reed *et al.* (2003).
[3]Feist *et al.* (2007).
[4]Archer *et al.* (2011).
[5]Determined from individual *in silico* models.
[6]Possible cellular compartments: cytoplasm, periplasm, extracellular space.
[7]FBA predicted growth rate on aerobic M9 minimal medium with glucose uptake rate of 10 mmol $gDW^{-1}h^{-1}$.
[8]Number of FBA predicted lethal gene-knockouts in this condition.
[9]Number of FBA predicted condition independent lethal gene-knockouts: all exchange reactions are allowed.

point in the mutant flux space that is 'closest' to the FBA determined flux state of the wild type (Segre *et al.*, 2002). Mathematically, MOMA minimizes the Euclidean distance:

$$D(w,x)=\sqrt{\sum_{i=1}^{N}(w_i-x_i)^2} \tag{2.6}$$

where w is the vector containing the value of the optimal wild-type fluxes as determined by FBA, and when $D(w,x)$ is minimal x is the MOMA predicted mutant fluxes. The growth rate corresponds to the biomass reaction flux, and is just one of the vector components of x (Segre *et al.*, 2002). While FBA is a linear optimization problem, the minimization of eqn 2.2 can be reformulated as a quadratic programming problem. Consequently, flux solutions to MOMA are always unique (Segre *et al.*, 2002).

The FBA and MOMA methods were compared by analysing their ability to (i) correctly predict growth phenotypes, and (ii) their flux predictions for central carbon metabolism were contrasted with experimental flux measurements (Emmerling *et al.*, 2002). The MOMA approach performed significantly better (Segre *et al.*, 2002). This should not come as a surprise, since the FBA method needs the mutant strains to be fully adapted to the growth conditions, which was not a part of the experimental protocol (Emmerling *et al.*, 2002).

ROOM prediction of mutant phenotypes

Regulatory on/off minimization (ROOM) (Shlomi *et al.*, 2005) is similar in approach to MOMA in that the idea is to identify a point in the mutant flux space that is, in a sense, 'close' to the wild-type FBA predicted flux state. While MOMA employs a Euclidean distance measure, where the contribution from each reaction is of equal weight, the idea behind ROOM is to minimize the number of significant flux changes as compared with the wild-type flux state (Shlomi *et al.*, 2005). It is reported that both FBA and ROOM are better than MOMA in predicting the final (adapted) metabolic growth phenotype, while MOMA is more accurate in predicting the initial (transient) growth rates experimentally observed after a gene knockout.

The ROOM approach is motivated by the assumption that regulatory changes leading to altered metabolic flux patterns come at an adaptation cost. Thus, mutants able to grow effectively in the early post-knockout phase are more likely to dominate the final population, and their flux distributions are close to that of the wild-type (Ideker *et al.*, 2001; Daran-Lapujade *et al.*, 2004; Shlomi *et al.*, 2005). ROOM is implemented using a mixed-integer linear programming (MILP) formulation, which allows for binary variables (being either 0 or 1) to lock the flux in any given reaction to a value close to the wild-type state. ROOM finds the final solution by minimizing the number of reactions that are not 'locked' into the wild-type state, while satisfying all the constraints of the FBA problem (Shlomi *et al.*, 2005).

Comparison of historical *E. coli* metabolic models

Comparison of model characteristics

The earliest models of *E. coli* central metabolism were published two decades ago (Majewski and Domach, 1990). It contained only 14 reactions and 17 metabolites. However, within eight years, it had been expanded to 317 reactions and 305 metabolites (Pramanik and Keasling, 1998). The first genome-scale metabolic model, designated iJE660, was published in 2000 and was based on the sequenced and annotated genome of *E. coli* K-12 strain MG1655 (Edwards and Palsson, 2000a). Here, the model naming convention is to use the initials of the model curator followed by the number of ORFs included. This model was further expanded (iJR904), and now the vast majority of the reactions were both elementally and charge balanced (Reed *et al.*, 2003). The third instalment of the *E. coli* K-12 strain MG1655 metabolic model, iAF1260, not only contains an increased number of metabolic genes, but the direction of the metabolic reactions were systematically checked for consistency with thermodynamic reversibility estimates (Feist *et al.*, 2007). Additionally, reactions in iAF1260 are assigned distinct cellular compartments: cytoplasm, periplasm, or extracellular space. Most recently, a genome-scale metabolic model has also been produced for *E. coli* W strain ATCC 9637, iCA1273 (Archer *et al.*, 2011). This was based on their sequencing and annotation of this strain, and the model was generated using various comparisons with iAF1260 and literature studies (Archer *et al.*, 2011). Of note: the biomass reaction in iCA1273 is the same as that suggested for iAF1260.

The first five lines in Table 2.1 provide a side-by-side comparison of the genome reconstructions' size characteristics: the number of reactions has increased almost three-fold, and the number of metabolites about 2.5 times while the number of accounted ORFs is almost doubled in the current *E. coli* models from that of iJE660.

Comparison of model predictions

While the four models, iJE660, iJR904, iAF1260, and iCA1273 cover from 660 to 1273 ORFs respectively, there are strong similarities in their simulated growth performance in simple media. An example of this is given in Table 2.1, where *in silico* optimal growth experiment was conducted on M9 minimal medium with glucose as the sole carbon source. For identical conditions, the models demonstrate almost identical growth rates. This should come as no surprise, since part of the model curation process is to test their growth performance (Edwards and Palsson, 2000a; Reed *et al.*, 2003; Feist *et al.*, 2007; Archer *et al.*, 2011). Important considerations in the curation process are (i) the flux value of the ATP maintenance reaction (Feist *et al.*, 2007) and (ii) the elemental composition and stoichiometric coefficients of the biomass reaction (Palsson, 2006; Feist *et al.*, 2007).

An indication of the difference between the models is shown in the last two lines in Table 2.1: whereas an *in silico* gene-knockout analysis of iJE660 in M9 medium with glucose as sole carbon source identifies 160 genes as being essential, a similar analysis of the largest model (iCA1273) identifies 191 genes as essential (Table 2.1). We also find varying levels of always essential genes (Almaas *et al.*, 2005) from an *in silico* gene knockout analysis in an environment where all transport reactions are allowed. In light of the extensive systematic improvements to model content, it is striking that the number of always essential genes in the models iJE660, iAF1270, and iCA1273 vary less than 9% from that predicted by iJE660. Here, iJR904 is an outlier by predicting 84 always essential genes.

In detailed, pair-wise model comparisons of iJR904 with iJE660 it was noticed that, while the model predictions agreed in a large number of cases, the improvements of iJR904 resulted in noticeable changes in model-predicted phenotype for some carbon sources under oxygen-limited conditions (Reed *et al.*, 2003). A systematic comparison of growth phenotypes in iAF1270 with iJR904 (Feist *et al.*, 2007) revealed that iAF1270 is a more robust metabolic reconstruction in that it is able to support growth in a significantly larger set of growth conditions using a wider variety of carbon, nitrogen, phosphorous, and sulfur sources (Feist *et al.*, 2007). In summary, iAF1270 (iJR904) correctly predicted *E. coli* ability or inability to grow in 76% (60%) of the cases when compared with experimental data (Feist *et al.*, 2007). Finally, the *E. coli* W strain model (iCA1273) is similar to iAF1270 in that, iCA1273 contains 61 genes excluded in iAF1270 and iAF1270 contains 48 genes excluded in iCA1273 (Archer *et al.*, 2011).

Biological tools for gene inactivation and control of expression levels

Bacteria, and particularly *E. coli*, have historically been used as major model organisms for exploration of fundamental genetic principles. Currently an enormous arsenal of technologies are available for genetic modification of these organisms, and although many of the available methods can be applied in diverse bacteria and even in complex eukaryotes, the procedures are generally much less laborious and time-consuming in *E. coli*. From a systems biology perspective it is a big advantage that *E. coli* (and most other bacteria) have many fewer genes than eukaryotes and can also multiply in synthetic media with a well-defined composition. These features also simplify the types of analyses described above.

In modelling it is important to be assisted by practical laboratory experiments, and in *E. coli* techniques for gene manipulation have been available for a long time. However, the methods have been continuously modified so that the desired modifications can now be done much faster. The genetic mapping of the entire *E. coli* genome by interrupted conjugation experiments involving numerous mutants represents one of the historical landmarks in these developments. Mutants were typically identified after random chemical mutagenesis followed by analysis of phenotypes by various and often innovative approaches. Access to newer technologies such as complete genome sequencing, polymerase chain reaction (PCR), and various array technologies have generally made gene modifications simpler and faster, and in the following sections we describe how one can now construct most desired gene modifications in the *E. coli* genome, and also in many other bacterial species.

Construction of single gene knock-outs at pre-selected sites

These methodologies have been available for a long time, and in the initial phases the methods were based on *recA*-dependent homologous recombination. A typical approach would be to first clone the chromosomal gene of interest (X) in a suitable plasmid vector, and then for example generate a deletion in X by the use of restriction endonucleases. The vector may be temperature-sensitive for replication or lack a protein required for replication. Such vectors can be propagated at the permissive temperature or in a specialized *E. coli* host that expresses the required replication protein. Examples of such hosts are S17-1 and S17-1 λpir, which can support replication of RK2 and R6K replicons, respectively (Simon *et al.*, 1983; de Lorenzo *et al.*, 1993). When these replicons are transferred to the host in which one wants to generate the corresponding deletion in the chromosome, they can become integrated at X by *recA*-dependent homologous recombination, and the recombinants are typically selected by an antibiotic resistance marker present in the vector. For temperature-sensitive replicons one also needs to select at high (non-permissive) temperature with respect to vector replication, while this is not required for replicons lacking an essential replication protein. Such vectors are generally described as suicide plasmids, and for manipulation of genes in hosts other than *E. coli* it is an advantage that most replicons designed for this host behave as suicide plasmids in other species.

The selected recombinants usually contain the entire vector integrated (single crossover), but in a second step excision and loss of the integrated vector may take place, leading to reconstruction of the original wild-type or a mutant carrying the desired mutation in X. The events leading to plasmid excision can be positively selected by also using another marker in the vector, such as the *Bacillus subtilis sacB* gene. Cells carrying the vector cannot grow in the presence of sucrose as carbon source (*sacB* expression is lethal), while cells in which the vector has been lost, can propagate (Gay *et al.*, 1985). The advantage of this type of approach is that it can be used in several bacterial species, as it is based on the hosts own recombination system, eliminating the need to ensure expression of a heterologous recombination system. In *E. coli* transformation of circular DNA is very efficient, while this is not always the case in other bacteria. If this is a problem gene transfer can usually be accomplished by conjugation (Pérez-Mendoza and de la Cruz, 2009). As an example, we have used such methods to generate both precise point mutations and a deletion in the chromosomal *algG* gene of *Pseudomonas fluorescens* (Gimmestad *et al.*, 2003). The methods will also in principle function on extrachromosomal genetic elements, such as bacterial artificial chromosomes (BAC). In general, the gene modification technologies that are based on homologous recombination are now often called recombineering, or recombinogenic engineering (Muyrers *et al.*, 2001; Sharan *et al.*, 2009).

The methods described above often work well, but also are associated with some major drawbacks. The dependence of the RecA function for recombination to take place leads to limitations in the choice of host strain that can be manipulated. This is because it is very common that laboratory strains of *E. coli* for various technical reasons are *recA* mutants. Secondly, the use of *in vitro* constructed plasmids as tools for the recombination events involves many and often time-consuming steps in the laboratory. The use of plasmids could previously not be avoided since most *E. coli* cells are not transformable with linear DNA, due to exonuclease activity that degrades the introduced DNA before recombination (Lorenz and Wackernagel, 1994).

The first major breakthrough representing a solution to the problems described above was reported by Zhang *et al.* (1998). They showed that linear PCR fragments containing short sequences homologous to the target could be recombined *in vivo*, and the target could be either a circular episome or the chromosome itself. The authors referred to their method as ET cloning because it was *recA*-independent and instead depends on the RecE and RecT functions, which originate from a cryptic prophage present in *E. coli* K12 strains (Clark *et al.*, 1993). These recombination functions are expressed in *sbcA* strains. The requirement for *sbcA* mutations could be circumvented by expressing RecE and RecT from a plasmid. For these methods to function the degradation of the introduced linear DNA must be prevented, and this can be achieved by using *recBC* strains. However, most *E. coli* strains used in experimental work are not *recBC*, and Zhang *et al.* (1998) showed that RecBC-mediated degradation of the introduced linear DNA could be prevented by transiently expressing the bacteriophage λ Red γ protein under control of the inducible P_{BAD} promoter. The same system was also used to control the level of *recE* expression, while RecT was expressed from a strong constitutive promoter. By using all these features many very important types of precise *in vivo* gene manipulations could be efficiently designed, including introduction of insertions, deletions and point mutations. The recombination procedures usually involve the insertion of an antibiotic resistance marker for selection, but it was also shown that such markers can be excised by using the site-specific yeast-derived FLP recombinase. The Cre/LoxP site-specific recombinase system originating from bacteriophage P1 is also frequently used for similar purposes (Fukiya *et al.*, 2004), and new such systems have been recently developed (Suzuki and Nakayama, 2011). The removal of the antibiotic resistance marker is important since it allows repeated gene manipulations to be carried out by using only a single marker, even though parts of the FLP or Cre target will not be removed. Furthermore, the recombinase used for this purpose can be located on a replication temperature-sensitive plasmid so that it can easily be eliminated after the desired gene modifications have been achieved.

The methods reported by Zhang *et al.* (1998) were later further developed by various technical modifications or improvements, based on the bacteriophage λ Red instead of the ET system, to control recombination (Datsenko and Wanner, 2000; Yu *et al.*, 2000). The λ Red system works well with very short homologous regions (below 40 nt), which even further simplifies the use of PCR, and an example of such a protocol is illustrated in Fig. 2.2. The same types of tools have been applied to generate large deletions in the *E. coli* chromosome (Fukiya *et al.*, 2004; Yu *et al.*, 2008) and the λ Red system has also been applied across species barriers in organisms such as *Pseudomonas aeruginosa*, *Pantotea anantis* and *Salmonella enterica* (Gerlach *et al.*, 2009; Katashkina *et al.*, 2009; Lesic and Rahme, 2008). Broad applications of the technologies may potentially face problems with expression of the recombineering functions developed originally for *E. coli*, and this has motivated searches for analogous systems in hosts of interest (Datta *et al.*, 2008). New systems have also been applied in medically important hosts such as mycobacteria (van Kessel and Hatfull, 2007).

Another popular type of tools are the so-called CRIM plasmids which are a large collection of conditionally replicating plasmids that can be integrated and excised from the *E. coli* chromosome by site-specific recombination at five different phage attachment sites (Haldimann and Wanner, 2001). These plasmids can carry any gene of interest and thus represent a simple set of tools for insertion of heterologous genes or homologous variant genes into the chromosome, usually leading to expression at physiologically relevant levels. The main conclusion to all these developments is that the most commonly desired targeted chromosomal (and episomal) gene manipulations can now be efficiently carried out in almost any *E. coli* strain by using linear PCR fragments, and many different and specific modifications can also be designed in any single strain, although this requires that the available procedures are repeated several times. All plasmid-based helper functions can be eliminated after completion of the procedures by taking advantage of their replication temperature-sensitive phenotypes.

Figure 2.2 (opposite) Graphical overview of a markerless deletion method based on λ-Red recombineering. In order to delete the target region on an *E. coli* chromosome, in this example between the region 'X' and 'Z', a linear PCR-generated DNA cassette is used. In the example illustrated this linear DNA contains a positive selection marker (Cm^R, resistance to chloramphenicol), a negative selection marker (*sacB*, blocks cell growth on sucrose), an *I-SceI* endonuclease recognition site (S) and three sequences (X, Y and Z) representing

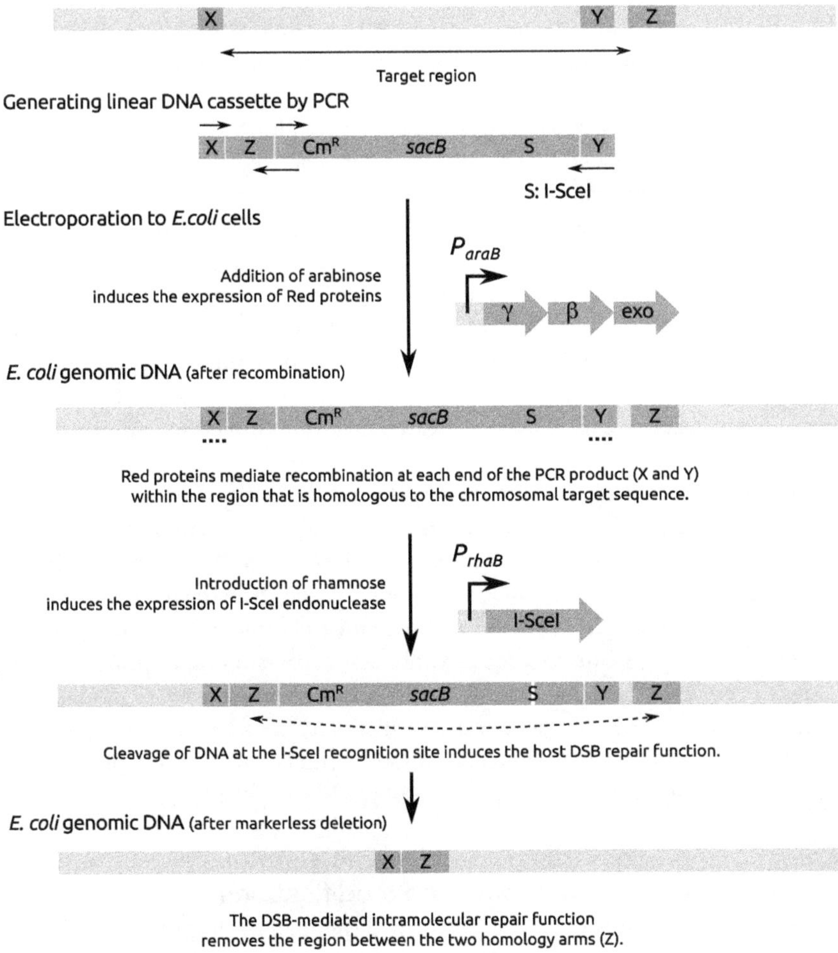

homology boxes to the target. The linear fragment spanning the region X-Y is generated by first PCR-amplifying two smaller fragments (0.5 and 3.0 kb, respectively). The 0.5 kb fragment contains X+Z and a small 5'-terminal part of the Cm^R marker. The source DNA for this is *E. coli* chromosomal DNA. The 3.0 kb fragment spans Cm^R-Y and is made from a plasmid (pSC1) already containing Cm^R, *sacB* and S. Finally, the two fragments are fused by recombinant PCR to generate the final linear DNA, taking advantage of the short overlap in the 5' part of Cm^R. This linear cassette is electroporated into *E. coli* cells containing plasmid pREDI, leading to integration of it into the chromosome, mediated by the λ-Red recombinase functions (gamma, beta and exo) encoded in pREDI, and controlled by the arabinose-inducible P_{araB} (P_{BAD}) promoter. At this stage host exonuclease activities are also inhibited by gamma, avoiding degradation of the introduced linear DNA. The integration event is selected by the Cm^R marker. In the next step expression of the *I-SceI* endonuclease gene (also located in pREDI) is triggered by activation of the rhamnose-inducible P_{rhaB} promoter, leading to a markersless deletion. The *I-SceI* endonuclease cleaves the chromosomal DNA at the *I-SceI* recognition site (S), and a markerless deletion will be generated via the host double-stranded DNA repair system. The *sacB* gene allows the elimination of any cells (by growth on sucrose) in which the *I-SceI* endonuclease mediated deletion has not taken place. Plasmid pREDI (providing the helper functions) can be eliminated at the end based on its temperature sensitivity for replication. Variations of this technology can also be used to generate other types of modifications, such as introduction of point mutations. The figure is a modification based on Yu *et al.* (2008).

Targeted knock-outs can also be achieved by a technology based on mobile group II introns, also called targetrons (Karberg *et al.*, 2001). These introns are retro-elements and insert themselves at specific DNA sequence targets with high frequency. Since the intron RNA is recognizing its target by base-pairing with its target DNA sequence the intron sequence can be modified such that the target can be changed to any desired gene. A software has been developed that can assist in the design of intron variants that can identify multiple potential targets in a single gene. The frequency of insertion is high enough to allow identification of insertion mutants without selection, but as this requires screening, the system has been further developed to include a selectable marker (Zhong *et al.*, 2003). With this method nearly 100% efficiency could be achieved, and the authors also describe a way by which the system could be used to assist in introduction of point mutations. The system can also be designed such that its specificity of insertion becomes more relaxed, and it has been shown that this can be used to create a more random type of inactivation library (Zhong *et al.*, 2003).

Like any other heterologous system, targetron (originates from a *Lactococcus lactis* intron) mobility is dependent on expression of the mobility functions, which potentially may limit the use of the system across species barriers. Yao and Lambowitz (2007) have shown that this problem can be minimized by coupling targetron to an inducible expression cassette (*xylS/Pm*) integrated into a broad-host-range plasmid RK2 mini-replicon, previously constructed in our laboratory for recombinant expression of, in principle, any gene. Major advantages of the *xylS/Pm* cassette is that it is known to be active in many bacterial species, that its inducer (benzoic acid or derivatives of it) does not require an active uptake system and that it can be gradually and continuously induced by varying the concentration of inducer molecules (Winther-Larsen *et al.*, 2000; Lale *et al.*, 2011).

Construction of random transposon insertion libraries and libraries of cells containing multiple random gene modifications

All the methods described above were primarily developed to simplify the inactivation of specific pre-selected genes, but alternatively one might make a library of mutants which investigators may use for example to screen for particular phenotypes. In *E. coli* such a collection has been made by creating a library (the Keio collection), which is designed in such a way that any out of 3985 *E. coli* genes are inactivated by in-frame deletions (Baba *et al.*, 2006). This latter point is very useful as it in most cases will eliminate polar effects of the transposon by inactivation of genes located in operons. The methodology used to construct this library is based on the λ Red system described above, and since each insertion point is known one can either select a specific mutant defined by a gene of interest, or one can screen the entire library for phenotypes of interest. Since the mutants can be arranged in a known order prior to screening any phenotype of interest is immediately identified with respect to the gene inactivated in that particular strain. Even though the λ Red technology is quite easy to use for any specific gene inactivation it is rather laborious to construct a library similar to that of the Keio collection in any given host. A simpler approach is to use random transposon insertions that can easily be constructed by transferring a suicide plasmid carrying a transposon like Tn*5* (often by conjugation) from *E. coli* to the host of interest (de Lorenzo *et al.*, 1990). This will yield a library with single insertions in each mutant. The disadvantage of this procedure is that it leads to polar effects for genes located in operons and that the insertion sites are unknown. However, they can be used to screen for phenotypes of interest, and the insertion sites are thereafter determined by direct sequencing of chromosomal DNA at the insert site, based on primers homologous to the transposon. It is also possible to identify the insertion sites prior to screening and thereby construct an arrayed library analogous to the Keio collection (but lacking in-frame inactivations). This type of technology is actually now becoming more popular owing to falling prices of DNA sequencing (Griffin *et al.*, 2011).

The methods described above yield libraries of cells carrying single insertions or mutations at sites that are different among each member of the library. For some purposes it may be useful to be able to efficiently generate multiple changes in single cells, and Williams *et al.* (2009) demonstrated that multiple traits can be exchanged between strains by simply using total DNA from a donor strain to transform a recipient recombineering strain. In another version of this type of idea complex mixtures of degenerate oligonucleotides were instead used to generate multiple modifications in a single cell (Wang *et al.*, 2009). An additional advantage of this approach is that the range of targets can be predetermined via the chosen design of the oligonucleotide mixture, and the authors described the use of this methodology for optimization of the metabolic flux through a biosynthetic pathway by mutagenizing ribosomal binding sites for selected genes in the pathway.

Methods for substitution of natural promoters by selected constitutive or regulated promoters

Most of the methods described above have been applied to generate various types of modifications affecting the coding sequences of genes, but Mnaimneh *et al.* (2004) developed a methodology which instead allowed control of the expression level of essential genes in yeast by inserting a regulated promoter upstream of each gene of interest. This idea has also been applied to *E. coli*. Mey *et al.* (2010) described the use of recombineering to control expression of a specific gene (*ppc*, encoding PEP carboxylase) by 'knocking-in' artificial promoter variants leading to different expression levels. In an extension of this type of approach Warner *et al.* (2010) reported a method called 'trackable multiplex recombineering' (TRMR) for use in *E. coli*. Their system involved the replacement of the endogenous promoters for most genes by the repressed but inducible $P_{LtetO\text{-}1}$ promoter. They made two sets of libraries involving so-called up and down versions of the promoter, where the down promoters are variants in which the ribosome binding site has been modified to reduce the efficiency of translation. All insertions were barcoded so that they could later be tracked by large-scale microarray analyses. The entire library population was then analysed with respect to their growth properties under various culture media conditions. These methodologies appear very attractive in that they allow such large scale studies within very short time frames, and therefore have the potential to generate very large and relevant datasets for use in systems biology with an acceptable input of resources and time.

Metabolomics and experimental determination of metabolic fluxes

The development of metabolomics and ^{13}C-labelled metabolic flux analysis have been driven by significant technological advancements the two last decades, in parallel with the advances in the other 'omics' fields: genomics, transcriptomics and proteomics. In this respect, metabolic flux analysis has also been given an 'omics' name – fluxomics (Wittmann,

2007). Metabolomics and fluxomics are analytically related as both focus on analysis of the metabolites in the cell, and can obviously, like gene knock-outs, generate data of critical importance for testing of models developed *in silico*. While Metabolomics aims at a qualitative or a quantitative analysis of the metabolite pool, experimental determination of the metabolic fluxes rely on the determination of the ^{13}C-isotope incorporation in the same metabolite pool by usage of ^{13}C-labelled substrate. Therefore, the same instrumental platforms are used for metabolomics and fluxomics, i.e. mass spectrometry (MS) and nuclear magnetic resonance (NMR). Here, a description of the current most important analytical techniques used in qualitative and quantitative analysis of metabolites as well as experimental determination of metabolic fluxes will be presented. The methodology is universal for all metabolite containing extracts and can therefore be used to study metabolite pools in bacteria, fungi and human cells and fluids.

Metabolomics

The strict definition of metabolomics is the global investigation of the metabolite pool in the biological system under study. More limited definitions are 'metabolite target analysis that is used for the quantitative analysis of a few metabolites', 'metabolite fingerprint analysis which is a high-throughput method used for the classification of a large number of samples' and 'metabolite profiling that is the most comprehensive analysis as it profiles one to several metabolite groups in a single run'. In addition, it is important to discriminate between targeted and non-targeted analysis. While the former focus only on known metabolites, the latter also includes analysis of unknown metabolites.

Metabolomics face particular analytical challenges due to the diverse physicochemical properties of the metabolites ranging from highly negatively charged organic acids and phosphometabolites to hydrophobic species such as fatty acids, steroids and pigments (e.g. carotenoids). They can also be present at a large range of concentrations, from a few molecules per cell to millimolar intracellular concentration. Both these aspects represent challenges from the analytical point of view and there is a common realization in the metabolomics community that no single analytical method can cover the whole metabolome, for example van der Werf and co-workers used six different analytical MS method for a comprehensive coverage of the *E. coli* metabolome (van der Werf *et al.*, 2007) and three MS methods for a comprehensive analysis of the *P. putida* S12 metabolome grown on different carbon sources (van der Werf *et al.*, 2008). Many protocols are already developed, nevertheless they can still be improved and new ones are added so that they readily can answer biological questions efficiently.

A metabolomics study comprises several steps, all of crucial importance for a successful outcome of the study. The first step is inactivation of cell metabolism, metabolite extraction and sample processing; the second step is the analysis itself which can be either a qualitative description of metabolites present in the extract or a true quantification of known metabolites, and the final step is the interpretation of the results. These three steps are discussed in the following sections.

Inactivation of cell metabolism, extraction and sample processing

The processing steps from taking a viable cell culture, inactivating the metabolism, and preserving the intracellular state during extraction and sample concentration have received massive attention. Largely, because it has been challenging to develop an optimal protocol, and the present status is that the perfect method still is not developed. The perfect method enables inactivation of metabolism within a fraction of a second, furthermore it must be possible to separate the cells from the extracellular medium without any loss of metabolites (i.e. a non-leakage situation), and finally all metabolites should be completely extracted and concentrated without any degradation. The latest method developed for yeast might probably be closest to the perfect method (Canelas *et al.*, 2008). This method is based on the pioneering method by de Koning and van Dam using quenching of cell metabolism by rapid transfer of cell culture to a –20°C water–methanol solution, followed by centrifugation and neutral extraction at the same low temperature (Dekoning and Vandam, 1992). The recent upgraded yeast method uses pure methanol quench solution and is technically challenging as it requires a centrifugation step at –40°C to prevent leakage of metabolites prior to the extraction steps. Bacteria, both Gram positive and negative, are unfortunately leaky in cold methanol solution. A decreased leakage was observed when replacing methanol with glycerol, but still some leakage remained (Villas-Boas and Bruheim, 2007). A commonly used solution has been the differential method implying simultaneous measurement of metabolites both leaked and extracted into supernatant with comparison to a cell-free sample (Winder *et al.*, 2008; Taymaz-Nikerel *et al.*, 2009). This is a challenging method from the analytical point of view as the background noise from the supernatant can be high. A third solution is fast filtration which must be considered a compromise as it takes 10–15 seconds from sampling start to transfer of the filter to an appropriate quenching solution. However, the subsequent processed extract enables the highest quality analysis since the cells on the filter can be washed several times to remove medium constituents. Extraction is most often performed at neutral pH, but acidic and alkaline extractions are also used. In this respect one should evaluate the pH stability of the metabolites of interest. There are many studies in bacteria and yeast addressing these issues by comparing the different methods, but they do not necessarily reach the same conclusion (Villas-Boas *et al.*, 2005; Bolten *et al.*, 2007; Spura *et al.*, 2009). It is therefore important for each individual project to evaluate which method is most appropriate for the current questions and problems raised.

Metabolite analysis

There are mainly two analytical platforms used for the analysis of metabolites: mass spectrometry (MS) and nuclear magnetic resonance (NMR) (Lindon and Nicholson, 2008). MS is especially popular due to its much higher sensitivity than NMR and due to its high resolving power when combined with a separation step. But, NMR is more of a complementary rather than a competitive technology, and it has proven its relevance for the analysis of metabolite pools. One advantage of NMR is that the analysis requires less processing of the sample as whole cells and tissues can be directly analysed. In addition, NMR is less discriminatory as all metabolites present at high concentrations can be detected while metabolite detection on a MS instrument is dependent upon MS instrument settings and separation conditions in the preceding chromatographic step. For example, negative ionization in MS is used for detection of negatively charged metabolites or metabolites that can be made negative during the ionization process, and likewise for positive ionization. The prerequisite for MS detection is that the metabolite is charged, alternatively it can be ionized, as the universal principle of MS is based on detection of mass/charge. Since a biological extract is far too complex to analyse by directly injecting it into a MS instrument, the metabolites in the extract must be separated in time prior to MS detection. Gas chromatography (GC) and high-pressure liquid chromatography (LC) are the two main separation techniques used for metabolomics applications, while capillary electrophoresis (CE) is a less used third alternative. The latter is due to significant lower sensitivity, but it has excellent separation properties for important metabolite classes as amino acids and nucleotides.

MS detectors have different properties depending on the mass detection principle they are derived from (Dettmer *et al.*, 2007; Bedair and Sumner, 2008). In general, a MS is either used for quantitative or for qualitative purposes, of which the former is the least expensive. The single (SQ) and triple quadrupole (QqQ) MS instruments are well recognized for their excellent properties in quantitative analysis. The SQ can be operated both in scan mode and in single ion monitoring (SIM), the latter mode used for quantitative analysis of known compounds as the MS only analyses pre-programmed mass/charges. The QqQ instrument has both higher sensitivity and selectivity as it can make use of unique fragmentation patterns arising in the collision cell in the MS. In multiple monitoring mode (MRM) the metabolites are first selected by the mass filter of Q1 (quadrupole 1) and then fragmented in Q2 (i.e. the collision cell) while Q3 works as a mass filter for the unique fragment ion(s) created from the precursor ion. One significant advantage with QqQ is that it can easily be combined with isotope dilution strategies for absolute quantification using ^{13}C-labelled standards (Wu *et al.*, 2005; Uehara *et al.*, 2009). The other available MS instruments are the more expensive ones having high resolution and high mass accuracy: time of flight (ToF), Orbitrap and Fourier transform ion cyclotron resonance (FT-ICR), the latter with a mass accuracy better than 1 ppm and resolution up to 2 million dependent on magnet strength. ToF-MS instruments benefit from high scan rates while the FT-MS types (Orbitrap and ICR) are slower working types, for example, the Orbitrap-MS must be operated in ToF resolution range (15,000 to 30,000) to be able to acquire 2–3 scans per second. However, they are attractive in metabolomics applications due to their stability and precision in accurate mass measurements. Accurate mass measurement in combination with MS/MS fragmentation analysis is a powerful tool in structural elucidation of metabolites, especially for the low-abundant compounds that are laborious to purify for NMR analysis. Lately, the introduction of an ion mobility-MS instrument (IMS) also makes it possible to separate and discriminate between structural isomer and other pairs of molecules with the same mass, for example, leucine and isoleucine, that is difficult to separate on GC and LC (Dwivedi *et al.*, 2008). The last MS detector type is the ion trap MS that has a MS^n capability of particular interest in structural elucidation, however it is also frequently used for quantitative purposes.

GC–MS has been regarded as the gold standard in metabolite analysis. It is a robust technique with very high separation efficiency (Fiehn, 2008). Importantly, the reproducible fragmentation occurring during electron impact (EI) ionization has enabled the building of large libraries of analyte mass spectra that are searchable (e.g. EPA/NIH/NIST-library containing over 170,000 entries and Wiley mass spectral database with 600,000 entries, and the metabolite specific Agilent-Fiehn library containing over 700 metabolites). The prerequisite for GC–MS analysis is that the analytes are thermally stable and volatile since the elution occurs in gas phase, usually with an increasing temperature (typical temperature range 50–300°C). Most metabolites must be made volatile through derivatization. The most common derivatization reactions are silylation, alkylation and acylation, and the choice of derivatization agent is based on the functional group of the metabolite of interest (Halket *et al.*, 2005). Silylation has been the most popular derivatization method for analysis of metabolites since it is a versatile derivatization reaction which covers many important metabolite classes such as sugars, alcohols, amines, amino acids, non-amino organic acids, aromatic compounds. Derivatization with silylation is often performed in two steps, where the first step is an oximation of carbonyl groups to oximes that stabilizes the reducing sugars in the open-ring formation and also prevent decarboxylation of α-ketoacids. Several metabolite profiling silyl-GC–MS studies have reported identification of over hundred metabolites, underlining the strength and potential of this methodology (Strelkov *et al.*, 2004; Koek *et al.*, 2006). Methyl chloroformate (MCF) derivatization has proven to be an alternative to silylation for amino acids and non-amino organic acids (Husek, 1998; Villas-Boas *et al.*, 2003). The advantage of the MCF method is that it is a rapid one step method that can be carried out in aqueous solution and the derivatives are extracted to a chloroform phase resulting in less contamination of the GC–MS. Recently, the MCF method has been further developed from an SQ scan method to a QqQ-MRM method enabling absolute quantification through the use of custom made deuterated MCF for introduction of an internal standard for all metabolites in the method (Kvitvang *et al.*, 2011).

LC–MS is a versatile and powerful analytical tool especially well suited for the analysis of non-volatile and thermally unstable compounds, and the mass detection range from low molecular metabolites to large biopolymers as proteins, polysaccharides and nucleic acids. The liquid mobile phase used for the LC separation step is evaporated in the electrospray ionization source (ESI) prior to MS detection that occurs under high vacuum conditions. Both quantitative and qualitative MS detectors are easily connected and used in combination with a LC separation. In general, LC–MS is more flexible than GC–MS with respect to change of ion source and columns as this switching can be done in minutes on a LC–MS system. LC–MS analysis of metabolites can roughly be divided in two large groups: those that can be separated using reverse phase (RP) chromatography and those that cannot. RP stationary phases (usually C_8 or C_{18}) are robust and reproducible phases that together with MS compatible volatile mobile phases as water, methanol, acetonitrile, and dichloromethane work excellent for non-charged and slightly hydrophobic analytes. For the most hydrophobic metabolites, for example, carotenoids, the atmospheric pressure chemical ionization (APCI) works better than the ESI source. The LC chromatographic challenge in Metabolomics has been for the highly charged metabolites that are not retained by standard RP elution conditions. Hence, these metabolite groups have attracted special attention, especially since important primary metabolites are highly charged. There are at present two LC–MS solutions to this challenge. The first is to add mobile phase modifiers, so called ion pair reagents, such as tributylamine, dibutylammonium, hexylamine, octylammonium (Coulier *et al.*, 2006; Luo *et al.*, 2007; Seifar *et al.*, 2009). These positively charged molecules mask the negative charge of the analytes and thereby improve the interaction between the hydrophobic RP stationary phase and the analytes. However, this approach has several disadvantages since the ion pair reagents impair on the ionization efficiency resulting in lower sensitivity of detection. In addition, they attach strongly to all surfaces in the LC–MS instrument so that extensive cleaning procedures must be applied when changing to a new analytical protocol. It is therefore preferable to have dedicated LC–MS instruments for ion pair chromatography. As an alternative to ion pair RP-LC methodology, hydrophilic interaction liquid chromatography (HILIC), has received most attention and has become increasingly popular for separation of polar analytes (Jandera, 2008; Cubbon *et al.*, 2010). HILIC is LC columns with silica surfaces that might or might not be slightly modified but still retain its normal phase (NP) behaviour. Common modifications include chemical binding to the silica gel with amino, amido, cyano, carbamate, diol and polyol ligands. While the traditional mobile phases for NP chromatography has been only organic solvents, the HILIC columns can be used with water. The usual mobile phase combination has been acetonitrile and water with ammonium and acetic acid to adjust the pH. Development of HILIC columns has significantly advanced among the column producers, especially since it has been experienced a rapid increasing market in bioanalysis. While these chromatographic systems earlier have suffered from poor stability and reproducibility, more and more reports are published with the use of HILIC technology showing that the chromatographic performance of HILIC columns have improved significantly. Bajad and co-workers tested seven different HILIC LC columns and could reliably detect over 140 metabolites for the column with best performance (Bajad *et al.*, 2006), while Schiesel *et al.* (2010) developed a quantitative zwitterionic ZIC-HILIC LC-QqQ MRM covering both central primary metabolites as well as secondary metabolites of β-lactam antibiotics synthesis including their precursors and extracellular degradation products.

NMR is a robust and very useful tool in structural elucidation of both small molecules and biopolymers. However, it has rather low sensitivity compared with MS (sensitivity limits around 1 μM for samples sizes around 500 μl). For the interested reader there are many excellent reviews on this topic, both covering microbial metabolomes and mammalian systems inclusive tissue analysis (Grivet *et al.*, 2003; Lindon and Nicholson, 2008; Wishart, 2008; Ludwig and Viant, 2010). The advantage of NMR is high selectivity and that it is non-destructive. Compared with MS it is not specific for any chemical species. ^{13}C- and ^{1}H-NMR is the standard nuclei used in NMR metabolomics. Typically proton NMR is used as ^{1}H-NMR is the most sensitive of the common nuclei and the analysis is rapid. ^{13}C-NMR has high information content because of large chemical shift dispersion and narrow lines that characterize this nucleus. ^{13}C has low abundance in natural compounds (approximately 1%) but sensitivity can be increased by using ^{13}C-enriched substrates. Microbial studies are benefitted from simple cultivations conditions where the carbon source (e.g. glucose) can be changed to the ^{13}C-labelled one (e.g. $^{13}C_6$-glucose). However, these analyses are expensive. Two-dimensional ^{1}H-^{13}C-NMR HSQC spectroscopy provides more information but is even less sensitive due to low abundance of the ^{13}C isotope, and ^{13}C enrichment from ^{13}C-added carbon source is highly recommended (Xi *et al.*, 2008). NMR can be used for quantification by comparison the NMR spectrum of the sample to a spectral reference library obtained from pure compounds using internal standard reference peaks (Wishart, 2008).

Interpretation of metabolome results

Interpretation of metabolomics results is in itself challenging, but the complete picture is not revealed until the metabolite data are integrated with other 'omics' data; hence this is a major challenge for systems biology (Kell, 2004; Nakahigashi *et al.*, 2009; Zhang *et al.*, 2010), as also discussed earlier in this chapter. If only Metabolomics results are to be interpreted, a common initial test of the data sets is to use chemometric tools to find and rank masses of importance (Goodacre *et al.*, 2004; Want, 2009). Usually the data sets relate to a set of perturbations (e.g. the comparison of different mutants, cultivation and stress conditions) and clustering analysis can be used to identify the metabolites which contribute most to the separation of the data. This approach can also be used for a non-target MS analysis with masses instead of known compounds. However, it can be a tedious puzzle to identify metabolites from a measured mass implying that a comprehensive interpretation of non-target MS data in a biological context can be very demanding. Visualization is often a very helpful tool in interpretation of data, so also for metabolite data. Several tools have been developed for the interactive exploration of metabolic pathways (Okuda *et al.*, 2008). They make use of available databases as KEGG, HMDB, but global models of metabolic networks developed for the particular organisms of interest also represent an excellent framework for representation and interpretation of metabolite data (Kell, 2004).

Experimental determination of metabolic fluxes

Metabolic fluxes are key determinants of the physiological state of a cell as it provides information about the cell metabolism complementary to gene expression, protein and metabolite pool compositions (Sauer, 2006). While a standard cultivation set-up with measurement of substrate consumption and product formation rates enables calculation of yields and productivities, this renders the interior of the cell almost as a black box. During the 1980s and 1990s the concept of stoichiometric metabolic flux analysis developed to provide information about intracellular metabolic flux distributions (Holms, 2001), and during the last decade developed further to comprise comprehensive global models (see 'Computational modelling of *E. coli* metabolism' section in this chapter). A major limitation of this approach is that the calculated solution only is found within the theoretical model used, for example, both the glycolytic pathway and Entner–Doudoroff pathway can be used in the same cell but stoichiometric metabolic analysis is not able to discriminate between these two pathways (Gunnarsson *et al.*, 2004). Furthermore, if incorrect enzyme NADPH and/or NADH dependencies are used in the stoichiometric model this will lead to incorrectly calculated metabolic flux distributions (Bruheim *et al.*, 2002). Therefore, the use of extra experimental information from isotope tracer experimentation has been developed to ensure more probable estimates of intracellular metabolic flux distributions. While the earliest experiments used radiorespirometry, for example Wang and co-workers used glucose labelled with ^{14}C in various positions and estimated that *E. coli* channelled about 28% of the glucose in the pentose phosphate pathway and 72% into the glycolytic pathway as early as in 1958 (Wang *et al.*, 1958), the presently used isotope is ^{13}C and labelling enrichment is analysed either with MS or NMR. ^{13}C-isotope metabolic flux analysis has contributed to elucidation of metabolic pathway activities, both to the discovery of new pathways and to optimize flux distributions through metabolic engineering. However, the resolution of current methodology limits flux estimations to the major and central metabolic pathways in a cell. The first applications of ^{13}C metabolic flux analysis made use of labelling data of proteinogenic amino acids harvested at isotopic steady-state situations. This is the simplest situation and least expensive as small cultivation volumes are required. Later, with improvement in precision and sensitivity in analytical instrumentation it has become routine to measure label incorporation directly in free soluble metabolite pool. This makes possible metabolic flux analysis of stimulus response experiments at non-stationary conditions (Antoniewicz *et al.*, 2007b; Wahl *et al.*, 2008). The interested reader is encouraged to study recent in-depth reviews for the more comprehensive status of the field, including both the tools and novel biological insights gained by their usage (Sauer, 2006; Tang *et al.*, 2009; Zamboni and Sauer, 2009). The tools, both experimental and computational, are still improved, and the recent technological advances in resolution, accuracy and sensitivity of modern analytical instrumentation as MS and NMR have strongly improved ^{13}C-isotope metabolic flux analysis. Such an experiment can be separated in three steps: cultivation with labelled substrate, measurement of labelling pattern, and finally calculation of the metabolic fluxes using the experimental data together with a computer model. The three steps are discussed more thoroughly below.

Cultivation with ^{13}C-isotope-labelled substrate

Labelling experiments are most easily performed in microbial cells that can be cultivated under simple and defined conditions, preferable with only one carbon source, so that noise from co-metabolism of several carbon sources (i.e. combination of sugars and amino acids) is not complicating the interpretation of the label enrichment pattern. Most published ^{13}C-labelled experiments have used glucose, either labelled in one or all positions. 1 g of ^{13}C-glucose costs around $100 implying that these kinds of analyses can be expensive. The standard research fermenter size for both batch and continuous cultivations using non-expensive chemicals is 1–3 l with an operating volume of 0.5–2 l. This ensure both excellent reactor monitoring and control, as well as plenty of culture for sampling, even for excessive time-series parallel transcriptome, proteome and metabolome sampling (Nieselt *et al.*, 2010). Clearly, it will be expensive to fill up a reactor volume of 1 l with 10 g/l ^{13}C-glucose ($1000), especially since usually a low culture volume (5–50 ml) is enough for the labelling analysis. Total sampling volume is depending on the scope of the experiment, i.e. one sampling at steady-state conditions or time-series sampling, and if it is the proteinogenic amino acids or the free soluble amino acid pool to be analysed. Two alternative strategies can therefore be applied, the first is to down scale the cultivation to mini-reactor systems and the other is to use a mixture of labelled and naturally labelled substrate. Several mini-reactor systems with operating volume 1–10 ml are commercially available and labelling studies in 1 ml deep-well micro plates have also been reported (Fischer *et al.*, 2004). When using fully labelled carbon substrates, it is common to use a mixture containing 10–20% labelled compound, while it is common to use a higher percentage of one position labelled substrates (Dauner *et al.*, 2001).

Measurement of labelling pattern enrichment

MS and NMR are the two experimental platforms used in isotopomer analysis (i.e. isotopomer/isotopic isomers are isomers having the same number of each isotopic atom but differing in their positions). The benefit of MS is the much higher sensitivity but it gives only information about the mass isotopomer distributions, while the NMR can directly yield information of the position of the labelled carbon atoms in the metabolite. NMR was a common technique in the early stages of development of ^{13}C-flux analysis, but currently MS, especially GC–MS, is most frequently used.

The simplest and most frequently used analysis is measurement at isotopic steady-state conditions. The proteinogenic amino acids can be used for the isotopomer analysis under such conditions because they acquire the same labelling pattern as the precursor metabolites from which they are synthesized from. Several amino acids are lost during acid hydrolysis (cysteine and tryptophan are degraded and glutamine and asparagine are converted to glutamate and aspartate respectively). However, the 16 remaining amino acids will provide isotope labelling information of eight precursor metabolites of the primary metabolism (ribose-5-phosphate, erythrose-4-phosphate, phosphoenolpyruvate, 3-phosphoglycerate, pyruvate, acetyl-CoA, oxaloacetate, α-ketoglutarate). If the free soluble metabolite pool is analysed, then confirmatory analysis directly on some of the precursor metabolites can be beneficial. The most common analytical method is GC–MS analysis of silylated metabolites. While N-methy-N-(trimethylsilyl) trifluoroacetamide (MSTFA) is used as derivatization agent for metabolomics purposes, tert-butyldimethylsilyl (TBD) MS is preferably used in ^{13}C-labelling studies as the TBDMS derivatives of amino acids are much more stable than respective trimethylsilyl (TMS)-derivatives. The reason TBDMS is not used for Metabolomics is that the reagent does not enable silylation of sugars (Dauner *et al.*, 2001; Antoniewicz *et al.*, 2007b). Alternatively, methyl chloroformate derivatization can be used as this yields very stable derivates of both amino acids and non-amino organic acids (Wentzel *et al.*, 2011). Common for all analysis, is the importance of performing a thorough analysis as integration algorithms, overlapping fragments and sample concentration influence on the mass isotopomer abundance calculation (Antoniewicz *et al.*, 2007a; Matsuoka and Shimizu, 2010).

LC–MS analysis has also been used to determine mass isotopomer abundances. Van Winden and colleagues used an advanced instrumentation with high performance anion exchange chromatography and a post-column desalting cell replacing the sodium ions in the mobile phase with protons prior to MS detection (van Winden *et al.*, 2005). Thus, they were able to analyse the isotope enrichment directly on metabolites in the upper glycolytic pathway and in the pentose phosphate pathway. This instrumentation configuration is unavailable in most laboratories, but with the developments in HILIC technology, LC–MS will probably gain more favour in metabolic flux analysis as it does for measurement of metabolite pools.

Calculation of metabolic fluxes from experimental data

Metabolic fluxes cannot be directly read from the labelling enrichments, but must be derived through models based on both biochemical and isotope balancing. The experimental labelling data can be used to calculate flux ratios that are used to constrain the flux at important branch point. Furthermore, absolute flux distributions can be estimated by additional usage of measured uptake and excretion fluxes. In general due to the advanced modelling and computing necessary for these tasks, ^{13}C metabolic flux estimation has been limited to a few laboratories with both strong experimental and theoretical expertise. But, for the non-expert user, ^{13}C metabolic flux analysis has been possible since several software tools have been made available for free academic use. 13CFLUX is probably the most advanced one (new version for simulation of very large networks is currently developed) as it contains functions for modelling, simulation, design, evaluation and statistical analysis (Wiechert *et al.*, 2001). Other simpler alternatives include FiatFlux (Zamboni *et al.*, 2005) and OpenFLUX (Quek *et al.*, 2009). Several recent reviews cover in detail the theoretical framework for ^{13}C metabolic flux analysis and can be accessed for a thorough introduction to the field (Sauer, 2006; Wittmann, 2007).

Concluding remarks

In this review we have discussed metabolic networks in *E. coli* from an *in silico* modelling point of view and we have also reviewed related wet lab activities, specifically with respect to the possibilities for constructing mutants and to the status of metabolomics. The computational modelling of cellular metabolism has recently been focused on the application of constraint-based optimization approaches. The availability of sequenced and annotated genomes, recent advances in the unsupervised generation of high-quality genome-scale metabolic reconstructions, and software packages for computer simulations has made metabolic modelling widely available and highly relevant. Access to relevant mutants is a very useful tool for verification of models, and the current status is that almost any relevant type of mutant design is now possible in *E. coli*. It is particularly important that in recent years it has become technically much easier to carry out such experiments, and the new technologies are also now to some extent applicable in other types of bacteria. Concerning metabolimics the methodologies for quantification of the fifty to a hundred most abundant metabolites and the central metabolic fluxes appear robust and reliable. As such, these data represent a valuable contribution to the understanding of cellular metabolism. However, significant further developments are expected in the near future so that an even more comprehensive picture of metabolite pools can be obtained.

References

Almaas, E., Oltvai, Z.N., and Barabasi, A.L. (2005). The activity reaction core and plasticity of metabolic networks. PLoS Comput. Biol. *1*, e68.

Antoniewicz, M.R., Kelleher, J.K., and Stephanopoulos, G. (2007a). Accurate assessment of amino acid mass isotopomer distributions for metabolic flux analysis. Anal. Chem. *79*, 7554–7559.

Antoniewicz, M.R., Kraynie, D.F., Laffend, L.A., Gonzalez-Lergier, J., Kelleher, J.K., and Stephanopoulos, G. (2007b). Metabolic flux analysis in a nonstationary system: Fed-batch fermentation of a high yielding strain of *E. coli* producing 1,3-propanediol. Metab. Eng. *9*, 277–292.

Archer, C., Kim, J., Jeong, H., Park, J., Vickers, C., Lee, S., and Nielsen, L. (2011). The genome sequence of *E. coli* W (ATCC 9637): comparative genome analysis and an improved genome-scale reconstruction of *E. coli*. BMC Genomics *12*, 9.

Baba, T., Ara, T., Hasegawa, M., Takai, Y., Okumura, Y., Baba, M., Datsenko, K.A., Tomita, M., Wanner, B.L., and Mori, H. (2006). Construction of *Escherichia coli* K-12 in-frame, single-gene knockout mutants: the Keio collection. Mol. Syst. Biol. *2*, 2006.0008.

Bajad, S.U., Lu, W.Y., Kimball, E.H., Yuan, J., Peterson, C., and Rabinowitz, J.D. (2006). Separation and quantitation of water soluble cellular metabolites by hydrophilic interaction chromatography-tandem mass spectrometry. J. Chromatogr. A *1125*, 76–88.

Becker, S.A., Feist, A.M., Mo, M.L., Hannum, G., Palsson, B.O., and Herrgard, M.J. (2007). Quantitative prediction of cellular metabolism with constraint-based models: the COBRA Toolbox. Nat. Protoc. *2*, 727–738.

Bedair, M., and Sumner, L.W. (2008). Current and emerging mass-spectrometry technologies for metabolomics. Trends Anal. Chem. *27*, 238–250.

Bolten, C.J., Kiefer, P., Letisse, F., Portais, J.C., and Wittmann, C. (2007). Sampling for metabolome analysis of microorganisms. Anal. Chem. *79*, 3843–3849.

Bruheim, P., Butler, M., and Ellingsen, T.E. (2002). A theoretical analysis of the biosynthesis of actinorhodin in a hyper-producing *Streptomyces lividans* strain cultivated on various carbon sources. Appl. Microbiol. Biotechnol. *58*, 735–742.

Canelas, A.B., Ras, C., ten Pierick, A., van Dam, J.C., Heijnen, J.J., and Van Gulik, W.M. (2008). Leakage-free rapid quenching technique for yeast metabolomics. Metabolomics *4*, 226–239.

Clark, A.J., Sharma, V., Brenowitz, S., Chu, C.C., Sandler, S., Satin, L., Templin, A., Berger, I., and Cohen, A. (1993). Genetic and molecular analyses of the C-terminal region of the *recE* gene from the Rac prophage of *Escherichia coli* K-12 reveal the *recT* gene. J. Bacteriol. *175*, 7673–7682.

Coulier, L., Bas, R., Jespersen, S., Verheij, E., van der Werf, M.J., and Hankemeier, T. (2006). Simultaneous quantitative analysis of metabolites using ion-pair liquid chromatography–electrospray ionization mass spectrometry. Anal. Chem. *78*, 6573–6582.

Covert, M.W., Knight, E.M., Reed, J.L., Herrgard, M.J., and Palsson, B.O. (2004). Integrating high-throughput and computational data elucidates bacterial networks. Nature *429*, 92–96.

Cubbon, S., Antonio, C., Wilson, J., and Thomas-Oates, J. (2010). Metabolomic applications of HILIC-LC–MS. Mass Spectrom. Rev. *29*, 671–684.

Daran-Lapujade, P., Jansen, M.L.A., Daran, J.M., van Gulik, W., de Winde, J.H., and Pronk, J.T. (2004). Role of transcriptional regulation in controlling fluxes in central carbon metabolism of *Saccharomyces cerevisiae* – A chemostat culture study. J. Biol. Chem. *279*, 9125-9138.

Datsenko, K.A., and Wanner, B.L. (2000). One-step inactivation of chromosomal genes in *Escherichia coli* K-12 using PCR products. Proc. Natl. Acad. Sci. U.S.A. *97*, 6640–6645.

Datta, S., Costantino, N., Zhou, X., and Court, D.L. (2008). Identification and analysis of recombineering functions from Gram-negative and Gram-positive bacteria and their phages. Proc. Natl. Acad. Sci. U.S.A. *105*, 1626–1631.

Dauner, M., Bailey, J.E., and Sauer, U. (2001). Metabolic flux analysis with a comprehensive isotopomer model in *Bacillus subtilis*. Biotechnol. Bioeng. *76*, 144–156.

Dekoning, W., and Vandam, K. (1992). A method for the determination of changes of glycolytic metabolites in yeast on a subsecond time scale using extraction at neutral pH. Anal. Biochem. *204*, 118–123.

Dettmer, K., Aronov, P.A., and Hammock, B.D. (2007). Mass spectrometry-based metabolomics. Mass Spectrum. Rev. *26*, 51–78.

Duarte, N.C., Herrgard, M.J., and Palsson, B.O. (2004). Reconstruction and validation of *Saccharomyces cerevisiae* iND750, a fully compartmentalized genome-scale metabolic model. Genome Res. *14*, 1298–1309.

Dwivedi, P., Wu, P., Klopsch, S.J., Puzon, G.J., Xun, L., and Hill, H.H. (2008). Metabolic profiling by ion mobility mass spectrometry (IMMS). Metabolomics *4*, 63–80.

Edwards, J.S., and Palsson, B.O. (2000a). The *Escherichia coli* MG1655 in silico metabolic genotype: Its definition, characteristics, and capabilities. Proc. Natl. Acad. Sci. U.S.A. *97*, 5528–5533.

Edwards, J.S., and Palsson, B.O. (2000b). Metabolic flux balance analysis and the in silico analysis of *Escherichia coli* K-12 gene deletions. BMC Bioinformatics *1*, 1.

Edwards, J.S., Ibarra, R.U., and Palsson, B.O. (2001). In silico predictions of *Escherichia coli* metabolic capabilities are consistent with experimental data. Nat. Biotechnol. *19*, 125–130.

Emmerling, M., Dauner, M., Ponti, A., Fiaux, J., Hochuli, M., Szyperski, T., Wuthrich, K., Bailey, J.E., and Sauer, U. (2002). Metabolic flux responses to pyruvate kinase knockout in *Escherichia coli*. J. Bacteriol. *184*, 152–164.

Feist, A.M., and Palsson, B.O. (2008). The growing scope of applications of genome-scale metabolic reconstructions using *Escherichia coli*. Nat. Biotechnol. *26*, 659–667.

Feist, A.M., and Palsson, B.O. (2010). The biomass objective function. Curr. Opin. Microbiol. *13*, 344–349.

Feist, A.M., Henry, C.S., Reed, J.L., Krummenacker, M., Joyce, A.R., Karp, P.D., Broadbelt, L.J., Hatzimanikatis, V., and Palsson, B.O. (2007). A genome-scale metabolic reconstruction for *Escherichia coli* K-12 MG1655 that accounts for 1260 ORFs and thermodynamic information. Mol. Syst. Biol. *3*, 121.

Feist, A.M., Herrgard, M.J., Thiele, I., Reed, J.L., and Palsson, B.O. (2009). Reconstruction of biochemical networks in microorganisms. Nat. Rev. Microbiol. 7, 129–143.

Fiehn, O. (2008). Extending the breadth of metabolite profiling by gas chromatography coupled to mass spectrometry. Trends Anal. Chem. 27, 261–269.

Fischer, E., Zamboni, N., and Sauer, U. (2004). High-throughput metabolic flux analysis based on gas chromatography-mass spectrometry derived C-13 constraints. Anal. Biochem. *325*, 308–316.

Fong, S.S., and Palsson, B.O. (2004). Metabolic gene-deletion strains of *Escherichia coli* evolve to computationally predicted growth phenotypes. Nat. Genet. *36*, 1056–1058.

Fukiya, S., Mizoguchi, H., and Mori, H. (2004). An improved method for deleting large regions of *Escherichia coli* K-12 chromosome using a combination of Cre/loxP and lambda Red. FEMS Microbiol. Lett. *234*, 325–331.

Gay, P., Coq, D.L., Steinmetz, M., Berkelman, T., and Kado, C.I. (1985). Positive selection procedure for entrapment of insertion sequence elements in gram-negative bacteria. J. Bacteriol. *164*, 918–921.

Gerlach, R.G., Jäckel, D., Hölzer, S.U., and Hensel, M. (2009). Rapid oligonucleotide-based recombineering of the chromosome of *Salmonella enterica*. Appl. Environ. Microbiol. 75, 1575–1580.

Gimmestad, M., Sletta, H., Ertesvåg, H., Bakkevig, K., Jain, S., Suh, S.J., Skjåk-Braek, G., Ellingsen, T.E., Ohman, D.E., and Valla, S. (2003). The *Pseudomonas fluorescens* AlgG protein, but not its mannuronan C-5-epimerase activity, is needed for alginate polymer formation. J. Bacteriol. *185*, 3515–3523.

Goodacre, R., Vaidyanathan, S., Dunn, W.B., Harrigan, G.G., and Kell, D.B. (2004). Metabolomics by numbers: acquiring and understanding global metabolite data. Trends Biotechnol. 22, 245–252.

Griffin, J.E., Gawronski, J.D., DeJesus, M.A., Ioerger, T.R., Akerley, B.J., and Sassetti, C.M. (2011). High-resolution phenotype profiling defines genes essential for mycobacterial growth and cholesterol catabolism. PLoS Pathog. 7, e1002251.

Grivet, J.P., Delort, A.M., and Portais, J.C. (2003). NMR and microbiology: from physiology to metabolomics. Biochimie *85*, 823–840.
Gunnarsson, N., Bruheim, P., and Nielsen, J. (2004). Glucose metabolism in the antibiotic producing actinomycete *Nonomuraea* sp ATCC 39727. Biotechnol. Bioeng. *88*, 652–663.
Haldimann, A., and Wanner, B.L. (2001). Conditional-replication, integration, excision, and retrieval plasmid-host systems for gene structure-function studies of bacteria. J. Bacteriol. *183*, 6384–6393.
Halket, J.M., Waterman, D., Przyborowska, A.M., Patel, R.K.P., Fraser, P.D., and Bramley, P.M. (2005). Chemical derivatization and mass spectral libraries in metabolic profiling by GC/MS and LC/MS/MS. J. Exp. Bot. *56*, 219–243.
Henry, C.S., DeJongh, M., Best, A.A., Frybarger, P.M., Linsay, B., and Stevens, R.L. (2010). High-throughput generation, optimization and analysis of genome-scale metabolic models. Nat. Biotechnol. *28*, 977–982.
Holms, H. (2001). Flux analysis: A basic tool of microbial physiology. Adv. Microb. Physiol. *45*, 271–340.
Hoppe, A., Hoffmann, S., Gerasch, A., Gille, C., and Holzhutter, H.-G. (2011). FASIMU: flexible software for flux-balance computation series in large metabolic networks. BMC Bioinformatics *12*, 28.
Hucka, M., Finney, A., Sauro, H.M., Bolouri, H., Doyle, J.C., Kitano, H., Arkin, A.P., Bornstein, B.J., Bray, D., Cornish-Bowden, A., *et al.* (2003). The systems biology markup language (SBML): a medium for representation and exchange of biochemical network models. Bioinformatics *19*, 524-531.
Husek, P. (1998). Chloroformates in gas chromatography as general purpose derivatizing agents. J. Chromatogr. B Analyt. Technol. Biomed. Life. Sci. *717*, 57–91.
Ibarra, R.U., Edwards, J.S., and Palsson, B.O. (2002). *Escherichia coli* K-12 undergoes adaptive evolution to achieve in silico predicted optimal growth. Nature *420*, 186–189.
Ideker, T., Thorsson, V., Ranish, J.A., Christmas, R., Buhler, J., Eng, J.K., Bumgarner, R., Goodlett, D.R., Aebersold, R., and Hood, L. (2001). Integrated genomic and proteomic analyses of a systematically perturbed metabolic network. Science *292*, 929-934.
Jandera, P. (2008). Stationary phases for hydrophilic interaction chromatography, their characterization and implementation into multidimensional chromatography concepts. J. Sep. Sci. *31*, 1421–1437.
Kanehisa, M., and Goto, S. (2000). KEGG: Kyoto Encyclopedia of Genes and Genomes. Nucleic Acids Res. *28*, 27–30.
Karberg, M., Guo, H., Zhong, J., Coon, R., Perutka, J., and Lambowitz, A.M. (2001). Group II introns as controllable gene targeting vectors for genetic manipulation of bacteria. Nat. Biotechnol. *19*, 1162–1167.
Katashkina, J.I., Hara, Y., Golubeva, L.I., Andreeva, I.G., Kuvaeva, T.M., and Mashko, S.V. (2009). Use of the lambda Red-recombineering method for genetic engineering of *Pantoea ananatis*. BMC Mol. Biol. *10*, 34.
Kell, D.B. (2004). Metabolomics and systems biology: making sense of the soup. Curr. Opin. Microbiol. *7*, 296–307.
van Kessel, J.C., and Hatfull, G.F. (2007). Recombineering in *Mycobacterium tuberculosis*. Nat. Methods *4*, 147–152.
Knorr, A.L., Jain, R., and Srivastava, R. (2007). Bayesian-based selection of metabolic objective functions. Bioinformatics *23*, 351–357.
Koek, M.M., Muilwijk, B., and van der Werf, M.J. (2006). Microbial metabolomics with gas chromatography/mass spectrometry. Anal. Chem. *78*, 3839–3839.
Krieger, C.J., Zhang, P.F., Mueller, L.A., Wang, A., Paley, S., Arnaud, M., Pick, J., Rhee, S.Y., and Karp, P.D. (2004). MetaCyc: a multiorganism database of metabolic pathways and enzymes. Nucleic Acids Res. *32*, D438-D442.
Kulikova, T., Aldebert, P., Althorpe, N., Baker, W., Bates, K., Browne, P., van den Broek, A., Cochrane, G., Duggan, K., Eberhardt, R., *et al.* (2004). The EMBL nucleotide sequence database. Nucleic Acids Res. *32*, D27-D30.
Kvitvang, H.F., Andreassen, T., Adam, T., Villas-Boas, S.G., and Bruheim, P. (2011). Highly sensitive GC–MS/MS method for absolute quantitation of amino and non-amino organic acids. Anal. Chem. *83*, 2705–2711.
Lale, R., Berg, L., Stüttgen, F., Netzer, R., Stafsnes, M., Brautaset, T., Aune, T.E.V., and Valla, S. (2011). Continuous control of the flow in biochemical pathways through 5′-UTR sequence modifications in mRNA expressed from the broad-host-range *Pm* promoter. Appl. Environ. Microbiol. *77*, 2648–2655.
Lesic, B., and Rahme, L.G. (2008). Use of the lambda Red recombinase system to rapidly generate mutants in *Pseudomonas aeruginosa*. BMC Mol. Biol. *9*, 20.

Lindon, J.C., and Nicholson, J.K. (2008). Analytical technologies for metabonomics and metabolomics, and multi-omic information recovery. Trends Anal. Chem. *27*, 194–204.

Lorenz, M.G., and Wackernagel, W. (1994). Bacterial gene transfer by natural genetic transformation in the environment. Microbiol. Rev. *58*, 563–602.

de Lorenzo, V., Herrero, M., Jakubzik, U., and Timmis, K.N. (1990). Mini-Tn5 transposon derivatives for insertion mutagenesis, promoter probing, and chromosomal insertion of cloned DNA in gram-negative eubacteria. J. Bacteriol. *172*, 6568–6572.

de Lorenzo, V., Eltis, L., Kessler, B., and Timmis, K.N. (1993). Analysis of *Pseudomonas* gene products using *lacIq*/P*trp-lac* plasmids and transposons that confer conditional phenotypes. Gene *123*, 17–24.

Ludwig, C., and Viant, M.R. (2010). Two-dimensional J-resolved NMR Spectroscopy: review of a key methodology in the metabolomics toolbox. Phytochem. Anal. *21*, 22–32.

Luo, B., Groenke, K., Takors, R., Wandrey, C., and Oldiges, M. (2007). Simultaneous determination of multiple intracellular metabolites in glycolysis, pentose phosphate pathway and tricarboxylic acid cycle by liquid chromatography-mass spectrometry. J. Chromatogr. A *1147*, 153–164.

Maglott, D., Ostell, J., Pruitt, K.D., and Tatusova, T. (2007). Entrez Gene: gene-centered information at NCBI. Nucleic Acids Res. *35*, D26-D31.

Majewski, R.A., and Domach, M.M. (1990). Simple constrained-optimization view of acetate overflow in *Escherichia coli*. Biotechnol. Bioeng. *35*, 732–738.

Matsuoka, Y., and Shimizu, K. (2010). The relationships between the metabolic fluxes and C-13-labeled isotopomer distribution for the flux analysis of the main metabolic pathways. Biochem. Eng. J. *49*, 326–336.

Mey, M.D., Maertens, J., Boogmans, S., Soetaert, W.K., Vandamme, E.J., Cunin, R., and Foulquié-Moreno, M.R. (2010). Promoter knock-in: a novel rational method for the fine tuning of genes. BMC Biotechnol. *10*, 26.

Mnaimneh, S., Davierwala, A.P., Haynes, J., Moffat, J., Peng, W.-T., Zhang, W., Yang, X., Pootoolal, J., Chua, G., Lopez, A., *et al.* (2004). Exploration of essential gene functions via titratable promoter alleles. Cell *118*, 31–44.

Muyrers, J.P., Zhang, Y., and Stewart, A.F. (2001). Techniques: Recombinogenic engineering-new options for cloning and manipulating DNA. Trends Biochem. Sci. *26*, 325–331.

Nakahigashi, K., Toya, Y., Ishii, N., Soga, T., Hasegawa, M., Watanabe, H., Takai, Y., Honma, M., Mori, H., and Tomita, M. (2009). Systematic phenome analysis of *Escherichia coli* multiple-knockout mutants reveals hidden reactions in central carbon metabolism. Mol. Syst. Biol. *5*, 306.

Nieselt, K., Battke, F., Herbig, A., Bruheim, P., Wentzel, A., Jakobsen, O.M., Sletta, H., Alam, M.T., Merlo, M.E., Moore, J., *et al.* (2010). The dynamic architecture of the metabolic switch in *Streptomyces coelicolor*. BMC Genomics *11*, 10.

Oberhardt, M.A., Palsson, B.O., and Papin, J.A. (2009). Applications of genome-scale metabolic reconstructions. Mol. Syst. Biol. *5*, 320.

Okuda, S., Yamada, T., Hamajima, M., Itoh, M., Katayama, T., Bork, P., Goto, S., and Kanehisa, M. (2008). KEGG Atlas mapping for global analysis of metabolic pathways. Nucleic Acids Res. *36*, W423-W426.

Orth, J.D., Thiele, I., and Palsson, B.O. (2010). What is flux balance analysis? Nat. Biotechnol. *28*, 245-248.

Ow, D.S.W., Lee, D.Y., Yap, M.G.S., and Oh, S.K.W. (2009). Identification of cellular objective for elucidating the physiological state of plasmid-bearing *Escherichia coli* using genome-scale in silico analysis. Biotechnol. Prog. *25*, 61–67.

Palsson, B. (2006). Systems Biology: Properties of Reconstructed Networks (Cambridge; New York, Cambridge University Press).

Pérez-Mendoza, D., and de la Cruz, F. (2009). *Escherichia coli* genes affecting recipient ability in plasmid conjugation: are there any? BMC Genomics *10*, 71.

Pramanik, J., and Keasling, J.D. (1998). Effect of *Escherichia coli* biomass composition on central metabolic fluxes predicted by a stoichiometric model. Biotechnol. Bioeng. *60*, 230–238.

Quek, L.E., Wittmann, C., Nielsen, L.K., and Kromer, J.O. (2009). OpenFLUX: efficient modelling software for C-13-based metabolic flux analysis. Microbiol. Cell Fact. *8*, 25.

Reed, J.L., and Palsson, B.O. (2003). Thirteen years of building constraint-based in silico models of *Escherichia coli*. J. Bacteriol. *185*, 2692–2699.

Reed, J.L., Vo, T.D., Schilling, C.H., and Palsson, B.O. (2003). An expanded genome-scale model of *Escherichia coli* K-12 (iJR904 GSM/GPR). Genome Biol. *4*, R54.

Rocha, I., Maia, P., Evangelista, P., Vilaca, P., Soares, S., Pinto, J.P., Nielsen, J., Patil, K.R., Ferreira, E.C., and Rocha, M. (2010). OptFlux: an open-source software platform for in silico metabolic engineering. BMC Syst. Biol. *4*, 45.

Sauer, U. (2006). Metabolic networks in motion: C-13-based flux analysis. Mol. Syst. Biol. *2*, 62.

Schiesel, S., Lammerhofer, M., and Lindner, W. (2010). Multitarget quantitative metabolic profiling of hydrophilic metabolites in fermentation broths of beta-lactam antibiotics production by HILIC-ESI-MS/MS. Anal. Bioanal. Chem. *396*, 1655–1679.

Schuetz, R., Kuepfer, L., and Sauer, U. (2007). Systematic evaluation of objective functions for predicting intracellular fluxes in *Escherichia coli*. Mol. Syst. Biol. *3*, 119.

Segre, D., Vitkup, D., and Church, G.M. (2002). Analysis of optimality in natural and perturbed metabolic networks. Proc. Natl. Acad. Sci. U.S.A. *99*, 15112–15117.

Seifar, R.M., Ras, C., van Dam, J.C., van Gulik, W.M., Heijnen, J.J., and van Winden, W.A. (2009). Simultaneous quantification of free nucleotides in complex biological samples using ion pair reversed phase liquid chromatography isotope dilution tandem mass spectrometry. Anal. Biochem. *388*, 213–219.

Sharan, S.K., Thomason, L.C., Kuznetsov, S.G., and Court, D.L. (2009). Recombineering: a homologous recombination-based method of genetic engineering. Nat. Protoc. *4*, 206–223.

Shlomi, T., Berkman, O., and Ruppin, E. (2005). Regulatory on/off minimization of metabolic flux changes after genetic perturbations. Proc. Natl. Acad. Sci. U.S.A. *102*, 7695–7700.

Shlomi, T., Cabili, M.N., Herrgard, M.J., Palsson, B.O., and Ruppin, E. (2008). Network-based prediction of human tissue-specific metabolism. Nat. Biotechnol. *26*, 1003–1010.

Simon, R., Priefer, U., and Pühler, A. (1983). A broad host range mobilization system for *in vivo* genetic engineering: transposon mutagenesis in Gram negative bacteria. Nat. Biotechnol. *1*, 784–791.

Spura, J., Reimer, L.C., Wieloch, P., Schreiber, K., Buchinger, S., and Schomburg, D. (2009). A method for enzyme quenching in microbial metabolome analysis successfully applied to gram-positive and gram-negative bacteria and yeast. Anal. Biochem. *394*, 192–201.

Strelkov, S., von Elstermann, M., and Schomburg, D. (2004). Comprehensive analysis of metabolites in *Corynebacterium glutamicum* by gas chromatography/mass spectrometry. Biol. Chem. *385*, 853–861.

Suzuki, E., and Nakayama, M. (2011). VCre/VloxP and SCre/SloxP: new site-specific recombination systems for genome engineering. Nucleic Acids Res. *39*, e49.

Tang, Y.J., Martin, H.G., Myers, S., Rodriguez, S., Baidoo, E.E.K., and Keasling, J.D. (2009). Advances in analysis of microbial metabolic fluxes via C-13 isotope labelling. Mass Spectrom. Rev. *28*, 362–375.

Taymaz-Nikerel, H., de Mey, M., Ras, C., ten Pierick, A., Seifar, R.M., Van Dam, J.C., Heijnen, J.J., and Van Glilik, W.M. (2009). Development and application of a differential method for reliable metabolome analysis in *Escherichia coli*. Anal. Biochem. *386*, 9–19.

Thiele, I., and Palsson, B.O. (2010). A protocol for generating a high-quality genome-scale metabolic reconstruction. Nat. Protoc. *5*, 93–121.

Uehara, T., Yokoi, A., Aoshima, K., Tanaka, S., Kadowaki, T., Tanaka, M., and Oda, Y. (2009). Quantitative phosphorus metabolomics using nanoflow liquid chromatography-tandem mass spectrometry and culture-derived comprehensive global internal standards. Anal. Chem. *81*, 3836–3842.

Villas-Boas, S.G., and Bruheim, P. (2007). Cold glycerol-saline: The promising quenching solution for accurate intracellular metabolite analysis of microbial cells. Anal. Biochem. *370*, 87–97.

Villas-Boas, S.G., Delicado, D.G., Akesson, M., and Nielsen, J. (2003). Simultaneous analysis of amino and nonamino organic acids as methyl chloroformate derivatives using gas chromatography-mass spectrometry. Anal. Biochem. *322*, 134–138.

Villas-Boas, S.G., Hojer-Pedersen, J., Akesson, M., Smedsgaard, J., and Nielsen, J. (2005). Global metabolite analysis of yeast: evaluation of sample preparation methods. Yeast *22*, 1155–1169.

Wahl, S.A., Noh, K., and Wiechert, W. (2008). C-13 labelling experiments at metabolic nonstationary conditions: An exploratory study. BMC Bioinformatics *9*, 152.

Wang, C.H., Stern, I., Gilmour, C.M., Klungsoyr, S., Reed, D.J., Bialy, J.J., Christensen, B.E., and Cheldelin, V.H. (1958). Comparative study of glucose catabolism by the radiorespirometric method. J. Bacteriol. *76*, 207–216.

Wang, H.H., Isaacs, F.J., Carr, P.A., Sun, Z.Z., Xu, G., Forest, C.R., and Church, G.M. (2009). Programming cells by multiplex genome engineering and accelerated evolution. Nature *460*, 894–898.

Want, E. (2009). Challenges in applying chemometrics to LC–MS-based global metabolite profile data. Bioanalysis *1*, 805–819.

Warner, J.R., Reeder, P.J., Karimpour-Fard, A., Woodruff, L.B.A., and Gill, R.T. (2010). Rapid profiling of a microbial genome using mixtures of barcoded oligonucleotides. Nat. Biotechnol. *28*, 856–862.

Wentzel, A., Overby, A., Bruheim, P., Jakobsen, O.M., Sletta, H., and Ellingsen, T.E. (2011). Optimized submerged batch fermentation strategy for systems scale studies of metabolic switching in *Streptomyces coelicolor*. (Submitted.)

van der Werf, M.J., Overkamp, K.M., Muilwijk, B., Coulier, L., and Hankemeier, T. (2007). Microbial metabolomics: Toward a platform with full metabolome coverage. Anal. Biochem. *370*, 17–25.

van der Werf, M.J., Overkamp, K.M., Muilwijk, B., Koek, M.M., van der Vat, B.J.C., Jellema, R.H., Coulier, L., and Hankemeier, T. (2008). Comprehensive analysis of the metabolome of *Pseudomonas putida* S12 grown on different carbon sources. Mol. Biosyst. *4*, 315–327.

Wiechert, W., Mollney, M., Petersen, S., and de Graaf, A.A. (2001). A universal framework for C-13 metabolic flux analysis. Metab. Eng. *3*, 265–283.

Williams, J.A., Luke, J., and Hodgson, C. (2009). Strain engineering by genome mass transfer: efficient chromosomal trait transfer method utilizing donor genomic DNA and recipient recombineering hosts. Mol. Biotechnol. *43*, 41–51.

van Winden, W.A., van Dam, J.C., Ras, C., Kleijn, R.J., Vinke, J.L., van Gulik, W.M., and Heijnen, J.J. (2005). Metabolic-flux analysis of *Saccharomyces cerevisiae* CEN.PK113–7D based on mass isotopomer measurements of C-13-labeled primary metabolites. FEMS Yeast Res. *5*, 559–568.

Winder, C.L., Dunn, W.B., Schuler, S., Broadhurst, D., Jarvis, R., Stephens, G.M., and Goodacre, R. (2008). Global metabolic profiling of *Escherichia coli* cultures: An evaluation of methods for quenching and extraction of intracellular metabolites. Anal. Chem. *80*, 2939–2948.

Winther-Larsen, H.C., Josefsen, K.D., Brautaset, T., and Valla, S. (2000). Parameters affecting gene expression from the *Pm* promoter in gram-negative bacteria. Metab. Eng. *2*, 79–91.

Wishart, D.S. (2008). Quantitative metabolomics using NMR. Trends Anal. Chem. *27*, 228–237.

Wittmann, C. (2007). Fluxome analysis using GC–MS. Microb. Cell Fact *6*, 6.

Wu, L., Mashego, M.R., van Dam, J.C., Proell, A.M., Vinke, J.L., Ras, C., van Winden, W.A., van Gulik, W.M., and Heijnen, J.J. (2005). Quantitative analysis of the microbial metabolome by isotope dilution mass spectrometry using uniformly C-13-labeled cell extracts as internal standards. Anal. Biochem. *336*, 164–171.

Xi, Y.X., de Ropp, J.S., Viant, M.R., Woodruff, D.L., and Yu, P. (2008). Improved identification of metabolites in complex mixtures using HSQC NMR spectroscopy. Anal. Chim. Acta *614*, 127–133.

Yao, J., and Lambowitz, A.M. (2007). Gene targeting in gram-negative bacteria by use of a mobile group II intron ("Targetron") expressed from a broad-host-range vector. Appl. Environ. Microbiol. *73*, 2735–2743.

Yu, B.J., Kang, K.H., Lee, J.H., Sung, B.H., Kim, M.S., and Kim, S.C. (2008). Rapid and efficient construction of markerless deletions in the *Escherichia coli* genome. Nucleic Acids Res. *36*, e84.

Yu, D., Ellis, H.M., Lee, E.C., Jenkins, N.A., Copeland, N.G., and Court, D.L. (2000). An efficient recombination system for chromosome engineering in *Escherichia coli*. Proc. Natl. Acad. Sci. U.S.A. *97*, 5978–5983.

Zamboni, N., and Sauer, U. (2009). Novel biological insights through metabolomics and C-13-flux analysis. Curr. Opin. Microbiol. *12*, 553–558.

Zamboni, N., Fischer, E., and Sauer, U. (2005). FiatFlux – a software for metabolic flux analysis from C-13-glucose experiments. BMC Bioinformatics *6*, 209.

Zhang, W.W., Li, F., and Nie, L. (2010). Integrating multiple 'omics' analysis for microbial biology: application and methodologies. Microbiology-Sgm *156*, 287–301.

Zhang, Y., Buchholz, F., Muyrers, J.P., and Stewart, A.F. (1998). A new logic for DNA engineering using recombination in *Escherichia coli*. Nat. Genet. *20*, 123–128.

Zhong, J., Karberg, M., and Lambowitz, A.M. (2003). Targeted and random bacterial gene disruption using a group II intron (targetron) vector containing a retrotransposition-activated selectable marker. Nucleic Acids Res. *31*, 1656–1664.

Bacterial Chemotaxis: Rising Complexity

3

Diana Clausznitzer, Judith P. Armitage and Robert G. Endres

Abstract

Bacterial chemotaxis is a paradigm for biological sensing and information transmission. The chemotaxis signal-transduction pathway allows cells to sense chemicals in their surroundings in order to regulate flagellated rotary motors, thus allowing them to swim towards nutrients and away from toxins. Importantly, cells are able to sense with remarkably high sensitivity over a wide range of chemical background concentrations. To make this possible, chemoreceptors do not signal independently but form clusters for amplification and integration of signals, as well as for adaptation to persistent stimulation. While chemotaxis in *Escherichia coli* has been exceptionally well characterized, new experimental facts still require revisions of existing models and thus further increase our understanding of sensing and signalling in bacteria. Additionally, experiments on other bacterial species such as *Bacillus subtilis* and *Rhodobacter sphaeroides* indicate that bacteria other than *E. coli* can have substantially different and more complex chemotaxis pathways, which provides renewed challenges for experimentalists and modellers alike. Here we discuss our current understanding as well as the frontiers of bacterial chemotaxis research.

Introduction

Sensory systems allow cells to receive stimuli from the environment and to respond accordingly. In bacterial chemotaxis, cells sense chemicals via receptors, which, when activated, signal to downstream components of the pathway. This ultimately regulates flagellated rotary motors and cell motility. Bacteria are too small to sense a spatial gradient, and therefore sense temporal changes in their environment (Dusenbery, 1998). To sense changes with time swimming bacteria must sense and briefly remember their environment, and modulate the frequency of direction changing when moving up or down the sensed gradient, resulting in a biased random walk (Berg, 1999). In the gram-negative species *Escherichia coli*, the best characterized bacterium (Sourjik, 2004; Wadhams and Armitage, 2004), sensing occurs with astonishingly high sensitivity. Cells are able to adapt their signalling and swimming behaviour to persistent stimulation by chemicals, allowing the sensory system to work over at least five orders of magnitude of chemical concentration (Berg and Brown, 1972; Mesibov *et al.*, 1973; Sourjik and Berg, 2002). Specifically, experiments show that cells respond to 3.2 nM ligand (Mao *et al.*, 2003), corresponding to only 3 molecules in the volume of the cell (assumed to be one femtolitre). Adaptation is precise to some attractant ligands such as aspartate, but not to others such as serine (Berg and Brown, 1972; Neumann *et al.* 2010).

Taken together, the bacterial chemosensory system exhibits exquisite sensing and signalling properties, providing a rich research topic for both experimentalists and modellers.

Bacterial cells are equipped with chemoreceptors, which enables them to respond to different chemoeffectors. These chemicals reflect the growth requirements and the environment of the specific species. *E. coli* has five different receptors (also called MCPs for methyl-accepting proteins), including the highly abundant major receptors Tar and Tsr for sensing the amino acids aspartate and serine, respectively (Li and Hazelbauer, 2004; Neumann *et al.* 2010). In contrast, in the gram-positive bacterium *Bacillus subtilis*, there are up to 40 different receptors and the magnetotactic bacterium, *Magnetospirillum magnetotacticum*, has over 50. The average number of receptor genes identified in bacterial genome sequences is about 15 (Alexander and Zhulin, 2007). Importantly, receptors cluster, predominantly at the cell poles, where clusters act as antenna to amplify weak signals (Maddock *et al.*, 1993; Bray *et al.* 1998; Thiem *et al.*, 2007; Endres, 2009). The clustering of receptors is not only conserved among bacteria, but also among Archaea (Gestwicki *et al.*, 2000). However, other bacterial species, such as *R. sphaeroides*, which are able to live under very different environmental conditions (Porter *et al.*, 2008), has an additional cytoplasmic chemotaxis-receptor cluster that potentially allows for sensing of the metabolic state (Thompson *et al.*, 2006). Indeed bioinformatic studies suggest over 50% bacterial species have more than one chemosensory pathway, and of those over 60% may be cytoplasmic.

Clusters do not signal as a coherent whole. There is indication that the units of signalling are smaller complexes of just a few receptors (Keymer *et al.*, 2006; Skoge *et al.*, 2006). Evidence stems from *in vivo* fluorescence resonance energy transfer (FRET) (Sourjik and Berg, 2002, 2004) measurements of signalling and super-resolution imaging by photo-activated localization microscopy (PALM) (Greenfield *et al.*, 2009). How cells integrate the different, and potentially opposing, chemical stimuli in receptor complexes, and even cell-external and internal signals is an important topic of current research.

The bacterial chemotaxis pathway is a member of so-called histidine kinase (HK), aspartate response regulator (RR) two-component systems (TCS). Most TCSs use a transmembrane HK protein which regulates to a diffusible RR. There are often hundreds of different TCSs in bacterial species, and even archaea, lower eukaryotes and plants, with the final output usually gene regulation or enzyme activity. These pathways regulate changes in cell physiology in response to environmental change, for example cell cycle, sporulation or virulence (Laub and Goulian, 2007; Wuichet *et al.*, 2010; Wuichet and Zhulin, 2010). The bacterial chemotaxis pathway is different from the canonical TCS in that the HK regulates the activity of two very different RRs, one regulating motor activity by protein–protein interaction and the other an enzyme resetting the signalling system.

E. coli contains most of the chemotaxis components found across bacteria and archaea. However, many of the auxiliary proteins found in other bacterial species are missing, making the pathway ideal for quantitative studies (see Fig. 3.1). In the absence of ligand binding, transmembrane receptors signal, via a conformational change to a cytoplasmic adapter protein CheW, and to the HK, CheA, which then autophosphorylates using ATP and passes the phosphoryl group to the RRs, CheY and CheB. Phosphorylated CheY-P diffuses to and binds to the motor and induces clockwise (CW) rotation, which promotes 'tumbling' and a new swim direction. In contrast, upon ligand binding, the kinase is inhibited and the concentration of CheY-P is reduced. As the affinity of CheY for the motor is lower, the motors rotate counter-clockwise (CCW) and the cell 'runs' smoothly. The excitation pathway needs

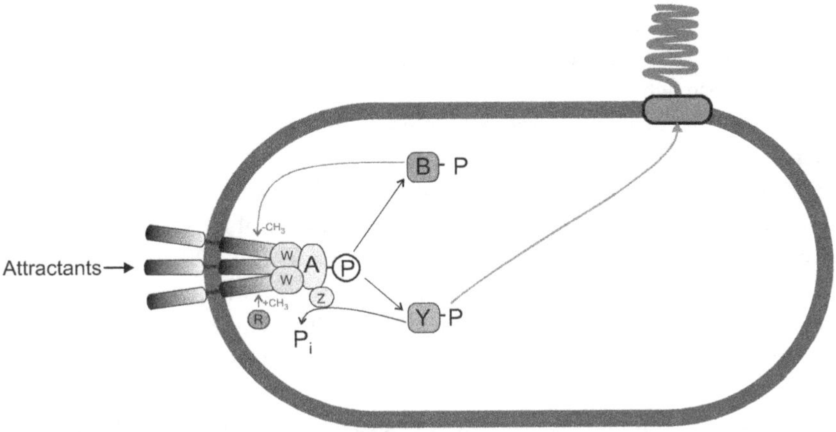

Figure 3.1 Schematic of *E. coli* chemotaxis pathway. Attractant binding to transmembrane chemoreceptors cause a conformational change that is transmitted to the cytoplasmic proteins. On reduction of attractant CheA (A), responding to linker CheW (W), transautophosphorylates, phosphoryl residues are transferred to CheY (Y), which binds to the flagellar switch to change the direction of rotation. CheY-P (Y-P) is dephosphorylated by CheZ (Z), releasing inorganic phosphate, Pi. CheA also phosphorylates the methyl esterase CheB (B), which with the constitutively active transferase, CheR (R) regulates the signalling state of the receptors, causing adaptation to the signal.

to terminate rapidly, to allow gradient sensing in a rapidly changing environment. *E. coli* has a dedicated phosphatase, CheZ, which dephosphorylates CheY-P and decreases the response time to addition of ligand.

This simple excitatory pathway would however not lead to the large observed dynamical range exhibited by bacteria. To avoid saturation of receptor signalling, receptors adapt via the slower process of receptor methylation. This depends on the methyl transferase CheR and demethylation by the methyl esterase, CheB. This type of covalent receptor modification is thought to influence the conformational state of the receptor, thus counter-acting the effect of attractant binding (Lai *et al.*, 2006b). While CheR is constitutively active, CheB needs activation by phosphorylation via CheA-P. As a result, if the receptor activity is too high, CheB is phosphorylated and demethylates receptors, reducing the activity of CheA. In contrast, if the activity is too low, CheR methylates receptors. Such negative feedback mechanisms are common in biological signalling.

Despite the strong similarity among the chemotaxis pathways in different bacterial species, there are also considerable differences (Szurmant and Ordal, 2004; Wuichet and Zhulin, 2010). Most chemotaxis pathways are considerably more complex than the *E. coli* pathway (Rao *et al.*, 2008; Porter *et al.*, 2008, 2011; Roberts *et al.*, 2010). In *B. subtilis* the phosphotransfer has the opposite effect in that ligand binding increases signalling activity and auto-phosphorylation of CheA, which promotes smooth swimming. Instead of CheZ, FliY, which localizes at the motors, has CheY-P phosphatase activity. Adaptation is also markedly different from *E. coli* (Rao *et al.*, 2008). First, the effect of methylation on the activity depends on which glutamate residue is methylated or demethylated. Second, in addition to CheR and CheB, there are proteins CheV, CheC and CheD, which form a second and a third adaptation system independent of receptor methylation. At least two out of the three are required for normal chemotaxis. CheV is an additional response regulator, which via its CheW-like domain, can influence CheA activity. CheC is a CheY phosphatase and CheD a receptor deamidase, which can either interact with the receptors or CheC, but only when bound to CheY-P. Hence, CheD–receptor interaction controls receptor activity and forms another adaptive feedback loop.

R. sphaeroides exhibits pathway complexity different again from *B. subtilis.* It has three complete chemosensory pathways, two of which are essential for chemotaxis under laboratory conditions (Porter and Armitage, 2002; Wadhams *et al.*, 2003; Porter *et al.*, 2008, 2011). These encode two different receptor clusters, one very much like that in *E. coli* and based around MCPs, the other in the cytoplasm with cytoplasmic receptors, together the two pathways control the bacterium's single flagellated motor (Fig. 3.2). Adaptation is implemented via two CheRs, one polar and one with the cytoplasmic cluster, and two diffuse CheBs; there is no CheV, CheD, and CheC, but termination is achieved by a phosphatase domain in the CheA HK of the cytoplasmic cluster. Unlike *E. coli,* the motor of *R. sphaeroides* rotates unidirectionally to produce smooth swimming. Directional changes occur when the flagellum is stopped by a molecular brake, and rotational diffusion reorients the cell. The pathway complexity may have arisen as a result of the complex life-style and metabolism of these bacteria. *R. sphaeroides* can grow by aerobic or anaerobic respiration, and can photosynthesize, fix carbon dioxide and fix nitrogen. Bioinformatic studies suggest that the two expressed pathways were introduced by horizontal gene transfer, with the native pathway only being expressed under a limited set of growth conditions. In addition to the usual attractants such as amino acids and sugars, cells also respond to organic acids, and

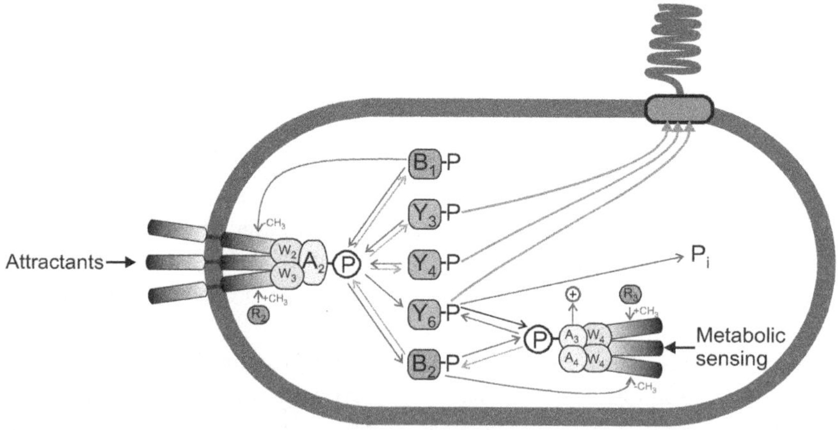

Figure 3.2 Schematic of *R. sphaeroides* chemotaxis pathway. As with *E. coli*, a transmembrane signalling cluster responds to changes in attractant binding, in this case with two CheW (W_2 and W_3) proteins regulating one CheA (A_2). CheA_2 can phosphotransfer to 3 CheY proteins (Y_3, Y_4 and Y_5) and two CheBs (B_1 and B_2). All CheYs can bind the motor, but only Y_6 can stop the motor. Adaptation involved a CheR (R_2) and a CheB (B_1). In addition a second chemosensory pathway is localized to mid-cell with soluble receptors. These signal through a localized CheW (W_4) to a CheA kinase/phosphatase split across two proteins (A_3 and A_4). $A_{3/4}$ can only phosphotransfer to Y_6 and B_2. Y_6-P is dephosphorylated by A_3. The diagram shows the interaction and phosphotransfers predicted from experiments and modelling.

chemotaxis towards some attractants requires their transport and uptake (Jeziore-Sassoon Y *et al.*, 1998).

While the act of chemosensing has fascinated researchers for well over a hundred years (essentially since the invention of the optical microscope) (Szurmant and Ordal, 2004), interest has increased since it has been found to have a fundamental role in processes such as symbiosis (Rawls *et al.*, 2007), virulence (Greer-Phillips *et al.*, 2004; Miller *et al.*, 2007; Williams *et al.*, 2007) and recently it has become a tool for developing synthetic biology (Kiessling *et al.*, 2006; Singh *et al.*, 2008; Kirby, 2010). Specifically, *E. coli* is a commensal bacterium, helping us to break down food and to produce essential vitamins. It may use chemotaxis to stay near the epithelial cell layer despite the large turnover of mucus (Gauger *et al*, 2007). At a more fundamental level, there is now a need in our understanding to go beyond single pathways and to study their more complex cousins, cross-talk with other pathways (Postma *et al.*, 1993; Hegde *et al.*, 2010), and context dependence, such as the influence of growth phase (Salman and Libchaber, 2007; Kalinin *et al.*, 2010). However, our main open question is whether we can use our extensive knowledge in *E. coli* chemotaxis to understand the complex chemotaxis pathway in *R. sphaeroides*. This would allow us to produce a general theory of bacterial chemotaxis. To approach this problem, we discuss our current understanding of both *E. coli* and *R. sphaeroides* pathways.

E. coli chemotaxis pathway

Signal amplification

Signal amplification has been one of the hallmarks of bacterial chemotaxis and subject of intense studies over the last 30 years. An increase in receptor occupancy by just 0.2% has been estimated to result in a 23% change in the bias of motor rotation (Segall *et al.*, 1986), indicating signal amplification (or gain) by a factor of about 100. One of the first theoretical models suggested an ordered array of allosterically coupled receptors (Bray *et al.*, 1998), which was subsequently formulated in terms of an Ising-type lattice model (Duke and Bray, 1999; Mello and Tu, 2003). However, with application of *in vivo* fluorescent resonance energy transfer (FRET) to *E. coli* chemotaxis (Sourjik and Berg, 2002, 2004), new detailed features of signal amplification came to light. Based on these, the Ising model was found insufficient to explain the signalling data of demethylated receptors (Skoge *et al.*, 2006). Instead, the Monod–Wyman–Changeux (MWC) model for allosteric coupling in protein complexes (Monod *et al.*, 1965) was revived and successfully applied to bacterial chemotaxis (Keymer *et al.*, 2006; Endres *et al.*, 2008).

In vivo FRET is generally used as a distance measure between donor and acceptor fluorophores fused to proteins of interest. The donor is excited by a laser and emits the light, unless near the acceptor, which, via energy transfer, can be excited and emit the light at a different (longer) wavelength. While the CheA kinase activity is most interesting to measure as it provides a direct readout of receptor signalling, an excellent FRET pair was found using response regulator CheY fused to yellow fluorescent protein (YFP) and phosphatase CheZ fused to cyan fluorescent protein (CFP) with CFP the donor and YFP the acceptor. FRET is nevertheless a measure of activity, since when CheA activity is high, there is a lot of CheY-P and hence CheY-P/CheZ pairs (for mathematical support, see Clausznitzer *et al.*, 2010). Based on the average FRET signal from many cells in a flow chamber, time-course data of chemotactic signalling to time-changing stimuli of ligand concentration as well as dose–response curves were measured (Sourjik and Berg, 2002, 2004). The latter can be obtained either for adapting cells, or for cells with receptors locked into specific adaptation-states.

In the MWC model, a small number N of receptors (about 10) form signalling complexes (see Fig. 3.3A for an illustration). Receptors are assumed to be either *on* (active) or *off* (inactive) (Asakura and Honda, 1984), and due to the strong receptor-receptor coupling, all receptors in a complex are also either on or off together (Keymer *et al.*, 2006). In this model, the activity of a complex is simply the probability to be on, given by

$$\text{Activity } A = P_{on} = \frac{1}{1+e^{\Delta F}} \tag{3.1}$$

where ΔF is the free energy difference between the on and the off states and, for a homogeneous receptor complex of a single receptor type, is simply N times the free energy difference between the on and off state of a single receptor Δf. This single-receptor free-energy difference is given by

$$\Delta f(c,m) = \Delta\varepsilon(m) + \log\left(\frac{1+c/K_D^{off}}{1+c/K_D^{on}}\right) \tag{3.2}$$

with c the ligand concentration, m the average receptor methylation level, and K_D^{on} and K_D^{off} the ligand dissociation constants of the on and off states, respectively. Energies are given in the unit of thermal energy K_BT.

The importance of the single parameter ΔF is a direct consequence of the two-state model. Implicitly, the free-energy difference is modulated by both ligand concentration and receptor methylation level. As ligand (attractant) turns receptors off to produce running cells, the off state binds ligand more strongly than the on state, requiring $K_D^{off} << K_D^{on}$. (The opposite holds for repellent chemicals, which are believed to favourably bind the on state

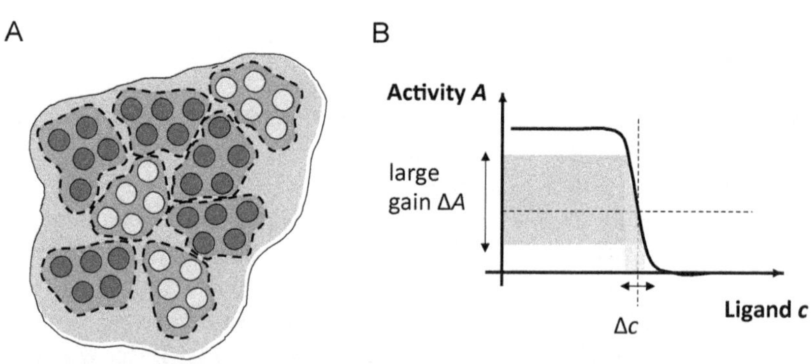

Figure 3.3 (A) Schematic of receptor cluster, which, within the MWC model, is assumed to be composed of smaller signalling complexes of strongly coupled receptors (here of five receptor dimers). Complexes are either fully on (light grey) or fully off (dark grey). (B) Illustration of amplification (gain) using a dose–response curve, mapping ligand concentration (input) onto receptor activity (output). A small stimulus (change in ligand concentration Δ*c*) leads to a large gain Δ*A*. In the MWC model, the gain is proportional to the number of receptors in the complex.

to increase cell tumbling.) As methylation is counteracting the effects of ligand binding, methylation is assumed to lower the on-state energy of the complex, facilitating the return to the correct adapted activity of runs and tumbles. The MWC model was successfully applied to explain signal amplification, signal integration, and precise adaptation (Endres and Wingreen, 2006; Keymer *et al.*, 2006; Endres *et al.*, 2008). These topics are explained in more detail below.

Fig. 3.3A illustrates the MWC model and Fig. 3.3B an exemplar dose–response curve showing how the ligand concentration is mapped onto receptor activity for a particular methylation level. Specifically, a small change in ligand concentration Δc relative to the background leads to a largely amplified change in activity ΔA relative to adapted value owing to receptor cooperation. In fact, this change in activity is proportional to the number of cognate receptors in the MWC complex.

Support for this model is provided by *in vivo* FRET measurements, recently conducted on cells expressing only the Tar receptor (Endres *et al.*, 2008). The ligand, α-methylaspartate (MeAsp) was used, which is an analogue of aspartate but cannot be metabolized and hence is suitable for chemotactic assays. Data in Fig. 3.4A (symbols) show widely different signalling behaviour of receptors locked into different adaptational states in line with the model (curves, see caption for details). Demethylated receptors are highly sensitive, i.e. turn off at tiny addition of ligand, and are largely inactive, even for low ligand concentrations (Keymer *et al.*, 2006). In contrast, receptors in higher adaptational states are fully active at low ligand concentrations and are very insensitive to ligand, i.e. turn off at very high ligand concentration (Keymer *et al.*, 2006). This concentration increases with increasing adaptational state, indicating increasingly desensitized receptors. Interestingly, Fig. 3.4B shows that the receptor cooperativity or complex size increases with adaptational state. This phenomenon was

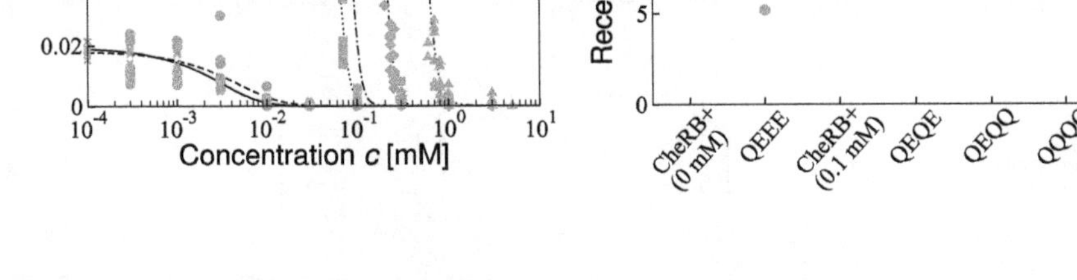

Figure 3.4 (A) Dose–response curves as measured by *in vivo* FRET (symbols) and corresponding fits by the MWC model (solid lines) for *E. coli* cells expressing only Tar receptors. Cell types include adapting (CheRB+) and non-adapting, engineered cheRcheB mutants (QEEE – circles, QEQE – squares, QEQQ – diamonds, and QQQQ triangles up with glutamate (E) or glutamine (Q) at four specific receptor modification sites). CheRB+ cells are adapted either to zero attractant (x symbols) or to 0.1 mmol MeAsp (+ symbols). (B) Corresponding receptor complex sizes with 95.4% confidence intervals, as extracted from the fitted MWC model. Data and model curves, fitted with Principal Component Analysis, are reproduced from Endres *et al.*, *2008.*

recently explained by studying the signal-to-noise ratio of receptor complexes (Aquino *et al.*, 2011). Since higher adaptational states correspond to higher ligand concentrations, high ligand concentrations exhibit less noise relative to background than low ligand concentrations (Berg and Purcell, 1977; Endres and Wingreen, 2008). Hence, larger complex sizes can be used to amplify weak signals more strongly without the detrimental effect of amplifying cell-external ligand noise.

In the MWC model described so far, receptors remain indefinitely associated with their complexes. This static model was recently extended to a more realistic dynamic version in which complexes of hexagonal lattice structure, as observed recently in cryo-electron microscopy (Briegel *et al.*, 2008, 2009), constantly assemble and disassemble (Hansen *et al.*, 2010). While the main findings of the static MWC remain, in the dynamic MWC model mainly cognate receptors become methylated instead of the uniform methylation of the whole complex in the static model. In the *E. coli* chemotaxis pathway, not only receptors exhibit cooperativity. The motor bias was measured to have a Hill coefficient of about 10, i.e. the motor switches abruptly as a function of CheY-P concentration (Cluzel *et al.*, 2000; Bai *et al.*, 2010). Hence, also the FliM-protein ring of the motor's rotor can approximately be described by an MWC-like model, but, due to the high temporal resolution achieved in recent imaging experiments, a conformation-spread model appears more suitable for explaining the motor-switching data (Bai *et al.*, 2010).

Signal integration

As the different types of receptors can sense specific ligands, the cell needs to somehow decide what stimulus to send to the motor. Based on the MWC model, it appears that such information is integrated at the receptor complex level. Complexes are believed to contain the different receptor types according to their abundances in the cell following multinomial distributions. This leads to a majority vote in order for the cell to decide to what ligand type to respond to predominantly, and whether to escape or to swim on. Fig. 3.5 illustrates that due to the strong receptor-receptor coupling, all receptors of mixed type are on or off together as in the homogeneous MWC model of a single receptor type. As a result there is only a single receptor-complex activity, which regulates the flagellated rotary motors.

As an example we consider receptor complexes composed of both Tar and Tsr receptors. The free energy difference of the receptor complex of N_a Tars and N_s Tsr receptors is given by:

$$\Delta F(c,m) = N\Delta\varepsilon(m) + N_a \log\left(\frac{1+c/K_a^{off}}{1+c/K_a^{on}}\right) + N_s \log\left(\frac{1+c/K_s^{off}}{1+c/K_s^{on}}\right) \tag{3.3}$$

with the complex size $N = N_a + N_s$. As there is only a single activity for the whole complex, the MWC model predicts that all receptors are methylated to an equal amount as indicated by the first term on the right-hand side of the equation. The second and third terms represent ligand binding to the Tar and Tsr receptors with their specific ligand dissociation constants, respectively. Owing to the prefactor, the ligand-specific change to the free energy difference is proportional to the number of cognate receptors in the complex. This free energy difference is also the larger the stronger the binding to ligand.

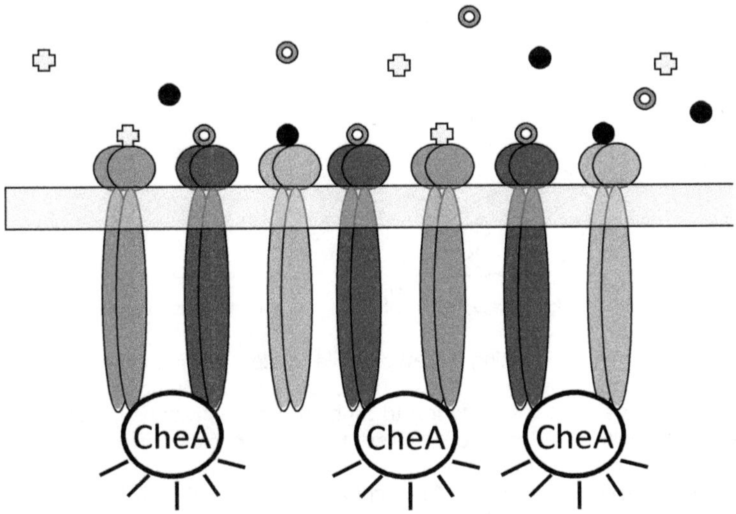

Figure 3.5 Schematic of a mixed receptor complex, containing multiple receptor types with different ligand specificity. In the MWC model, the signalling bias of each receptor type gets averaged and hence integrated according to the receptor-type abundance in the complex and their ligand-binding strength.

The majority vote has multiple implications. In principle, the opposing effects of an attractant and repellent could exactly cancel one another, which leads to neither swimming to the attractant nor escaping from the repellent. To ensure escape or protection, cells may rely on additional sensing mechanisms and signalling pathways. Receptors, which do not bind ligand, still contribute methylation sites and hence can contribute to exact adaptation of the receptor complex. As non-cognate receptors may bind ligand at higher concentrations, the dynamic range of mixed receptor complexes is potentially higher than that of homogeneous receptor complexes. This will be demonstrated next.

Fig. 3.6 shows *in vivo* FRET data and results from the MWC model of adapting wild-type cells, exposed to step-increases and decreases of ligand MeAsp. Fig. 3.6A shows the initial response amplitudes of adapted cells for four different ambient concentrations (for details see legend). The upper and lower left most curves show that receptors adapted to zero ambient are highly sensitive, and that with increasing ambient, receptors become increasingly insensitive. Importantly, the dynamic range is significantly increased to about five orders of magnitude compared with the Tar-only cells from Fig. 3.4. To see this more clearly, panel B shows the logarithmic sensitivity, defined by:

$$S = \frac{d\log A}{d\log c} \approx \frac{\Delta A/\bar{A}}{\Delta c/\bar{c}} \tag{3.4}$$

where the numerator is the relative change of the activity and the denominator is the relative change of the ligand concentration, set to 0.1 (10%) here. It is clear that while the major response is between $K_a^{\mathrm{off}} = 0.02$ mmol and $K_a^{\mathrm{on}} = 0.5$ mmol, the response continues further

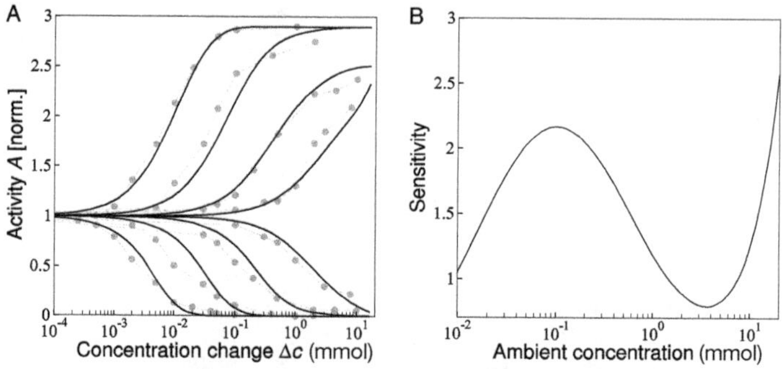

Figure 3.6 (A) Response of adapting wild-type cells to step changes of MeAsp concentration at different ambient concentrations. Dose–response curves: Symbols represent averaged response from FRET data after adaptation to ambient concentrations 0, 0.1, 0.5 and 2 mmol as measured by Sourjik and Berg (2002); filled and open circles correspond to response to addition and removal of attractant, respectively). Solid lines represent the dynamic MWC model of mixed Tar/Tsr-receptor complexes (Clausznitzer *et al., 2010*). (B) Logarithmic sensitivity from model as defined in the text. The peak at about 10^{-1} mmol of MeAsp corresponds to the response from Tar receptors. The rise at the right end of the curve shows the response of the Tsr receptors.

to higher concentrations due to MeAsp binding to Tsr with an estimated K_s^{off} of 100 mmol (Endres and Wingreen, 2006).

So far, we have discussed only ligands, such as amino acids, which bind directly to the highly abundant Tar and Tsr receptors in the periplasmic space. However, some ligands, such as sugars and dipeptides, first need to bind periplasmic-binding proteins, which then bind to the receptor for chemosensing or ATP-binding cassette (ABC) transporters for uptake. An exception is sugar maltose, which binds indirectly to the major receptor Tar. Recent work showed that such indirect binding allows cells to better control sensitivity to specific ligands in response to their nutrient environment, and to coordinate chemotaxis with ligand uptake (Neumann *et al.*, 2010). This occurs at the cost of a much narrower dynamic range in the case of indirect binding.

Robust adaptation

Adaptation of the frequency of tumbling was early found to be precise to aspartate over several orders of magnitude of ligand concentration (Berg and Brown, 1972). To explain precise adaptation, Barkai and Leibler put forward a model in which the rates of receptor methylation and demethylation only depend on the receptor activity but not substrate concentration, such as availability of methylation sites on the receptors, or the external ligand concentration (Barkai and Leibler, 1997). This type of enzyme specificity is illustrated in Fig. 3.7, where enzyme CheR can only bind to the inactive receptor conformation and CheB can only bind to the active receptor conformation in line with experimental observation (Toews *et al.*, 1979; Lai *et al.*, 2006a; Li and Hazelbauer, 2006). In mathematical terms, the rate of change of the average receptor methylation level may be given by

$$\frac{dm}{dt} = k_R(1-A) - k_B A \qquad (3.5)$$

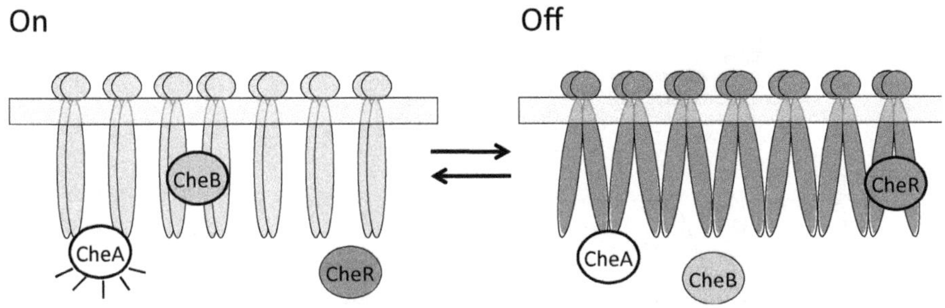

Figure 3.7 Schematic of receptor complex. Receptors in the on state (light grey) signal to the receptor bound kinase CheA to be active, and receptors in the off state (dark grey) inhibit the kinase. In a simple model for precise adaptation, receptors in the on state become demethylated by enzyme CheB, which lowers the signalling activity, and receptors in the off state become methylated by enzyme CheR, which increases the signalling activity.

where k_R is the rate constant of methylation by enzyme CheR of inactive receptors (to increase their activity) and k_B is the rate constant of demethylation by enzyme CheB (to decrease their activity). Technically, even CheB needs activation by phosphorylation, leading potentially to higher powers of the activity for demethylation in order to describe adaptation time-course data (Clausznitzer *et al.*, 2010). The Barkai and Leibler model was subsequently identified to be an example of integral feedback control known from engineering (Yi *et al.*, 2000): receptor methylation is not proportional to the current mismatch between the desired activity and the actual activity, but the time-integrated mismatch, leading to robust tracking of the correct adapted activity.

The implications of the Barkai and Leibler model have been tested numerously. While the adaptation time and value of the adapted activity depend on the enzyme and receptor expression levels, and hence vary from cell to cell, the precision of adaptation was indeed found to be a robust trait (Alon *et al.*, 1999). Since the number of adaptation enzymes is very small (a few hundred copies; Li and Hazelbauer, 2004), adaptation produces substantial low frequency noise in the motor (Korobkova *et al.*, 2004), which was hypothesized to lead to beneficial cell-to-cell variability (Emonent and Cluzel, 2008).

Despite the successes of the Barkai and Leibler model to explain robust, precise adaptation, stochastic simulations of adaptation of the MWC model surprisingly showed that MWC complexes do not adapt precisely (Endres and Wingreen, 2006). The reason for this stochastic effect is that the individual receptor methylation levels of the receptors in the complex do not correlate well with the activity of the whole complex. Hence, sometimes fully methylated receptors are attempted to be even further methylated or fully demethylated receptors are attempted to be even more demethylated. Since this does not work, such aborted attempts lead to methylation level-dependent rates and hence imprecise adaptation. This problem is remedied when considering that the enzymes CheR and CheB bind to the C-terminal tethers of the major receptors to act upon groups of 5–7 receptor dimers. These so-called *assistance neighbourhoods* (Li and Hazelbauer, 2005) restore precise adaptation as enzymes have more sites available for modification without running into the limitations of individual receptors (Endres and Wingreen, 2006).

More recently, the limitations of precise adaptation attracted renewed interest, as it became apparent that receptor signalling does not adapt perfectly (Fig. 3.8A, Clausznitzer

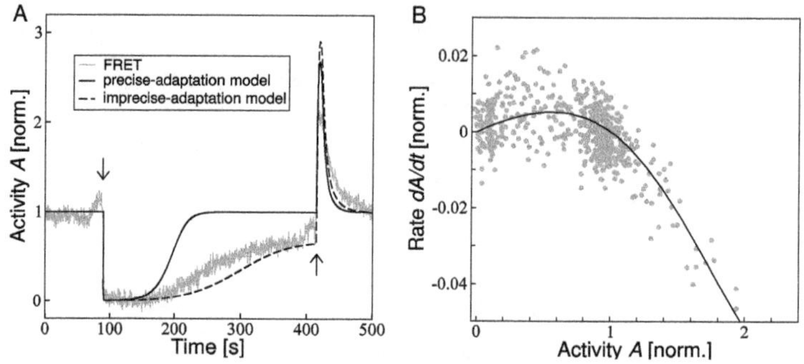

Figure 3.8 (A) Adaptation time course. Representative FRET data in response to a change in MeAsp concentration of 0.4 mmol relative to 0.1 mmol ambient (grey line). Models of precise (solid line) and imprecise (dashed line) adaptation are also shown. In the model for imprecise adaptation, rates of methylation and demethylation do not solely depend activity, but also on receptor methylation level. (B) Data collapse with rate of activity change as a function of activity. Both FRET data (grey symbols) and model (black curve) are shown. Data and curves reproduced from Clausznitzer *et al.* (2010).

et al., 2010; Meir *et al.*, 2010) – somewhat in contrast to earlier measurements at the cell behavioural level (Berg and Brown, 1972). A convenient way to quantify imprecise adaptation is to collapse all available time-course data onto a single curve irrespective of ambient or stimulus strength. This can be done by writing the above equation dm/dt for the rate of change of the methylation level in terms of the activity $dA/dt = dA/dm \cdot dm/dt = f(A)$. The resulting rate of change of the activity is just a function of the activity $f(A)$, and hence, if adaptation is precise, $dA/dt(A)$ extracted from time-course data are predicted to collapse onto $f(A)$. This is illustrated in Fig. 3.8B using FRET data (such as shown in Fig. 3.8A). Further analysis shows that there are substantial deviations from the predicted function of precise adaptation due to imprecision in adaptation for large ligand concentrations, as well as due to the dynamics of ligand in the flow chamber used in the experiments. The latter also contribute to changes in activity.

R. sphaeroides chemotaxis pathway

R. sphaeroides has two chemosensory pathways; one localizes to the cell poles with the MCPs, of which there are eight encoded randomly around the genome (Mackenzie *et al.*, 2001), and the other pathway localizes to the mid-cell with the cytoplasmic chemoreceptors, of which there are four, all encoded within chemotaxis operons. The polar chemosensory pathway has two CheWs, $CheW_2$ and $CheW_3$, associated along with $CheA_2$ and $CheR_2$. $CheB_1$, encoded in the operon expressing the polar cluster proteins appears to be diffuse in the cytoplasm, as does $CheB_2$, encoded with the cytoplasmic cluster proteins (reviewed in Porter *et al.* 2008b, 2011). The cytoplasmic cluster has, in addition to the cytoplasmic chemoreceptors, or Tlps, $CheW_4$ and a CheA divided across two proteins: $CheA_4$ has the kinase and dimerization domain while $CheA_3$ has the conserved histidine on an Hpt domain, plus a phosphatase domain specific for only one CheY-P, $CheY_6$-P (Porter *et al.*, 2008a). $CheA_4$ is phosphorylated by ATP and passes the phosphoryl to the conserved His on $CheA_3$, which then phosphorylates $CheY_6$ and $CheB_2$, and no other RR. $CheA_3$ also dephosphorylates $CheY_6$-P, but no other RR.

There are at least six CheY homologues encoded in the *R. sphaeroides* genome (Mackenzie *et al.*, 2001; Porter *et al.* 2008b). Three are associated with the operon not expressed under laboratory conditions, while $CheY_3$ is encoded with the polar proteins and $CheY_6$ with the cytoplasmic cluster. Chemotaxis requires $CheY_6$, as the only CheY able to stop the motor, but its activity appears modulated by either $CheY_3$ or $CheY_4$ (encoded with an MCP elsewhere on the genome; Porter *et al.*, 2006).

Both chemosensory pathways are essential for chemotaxis, and the inheritance of a cluster by a daughter cell on cell division is regulated by a ParA homologue, PpfA, encoded in the cytoplasmic cluster operon next to the cytoplasmic chemoreceptor, TlpT (Thompson *et al.*, 2006). The cytoplasmic cluster is usually localized at about mid cell in a newly divided cell and as the cell grows clusters move to ¼ and ¾ suggesting a localization system. Deletion of PpfA results in cluster formation, but not positioning, with only one cluster per cell, randomly positioned, so that on division one daughter lacks a cluster and the other inherits a large cluster. Deletion of TlpT results in the complete loss of cluster formation, with all chemotaxis cluster proteins being diffuse, suggesting TlpT acts as a scaffold for cluster formation, while PpfA actively positions the cluster at mid- or future mid-cell (M.A. Roberts and J.P. Armitage, unpublished). Interestingly the chemosensory proteins only function if they are in the correct cellular location. The proteins will all complement equivalent deletions in *E. coli* cells, but will not complement deletions of the equivalent protein in *R. sphaeroides*. Indeed, if the HK proteins are localized to the 'wrong' cluster by expression with the alternative receptor binding domain (e.g. $CheA_2$ has its MCP binding domain replaced with the Tlp binding domain of $CheA_4$), the proteins will incorporate into the other cluster, but do not support chemotaxis (Scott *et al.*, 2010). Surprisingly, even when highly over expressed in deletion backgrounds, the CheA proteins will not localize to the 'other' cluster, although they will localize to the *E. coli* MCP cluster, suggesting specificity is more highly regulated in *R. sphaeroides* and is essential for function.

In vitro analysis of phosphotransfer showed that $CheA_2$ from the polar cluster can phosphotransfer to all CheY and CheB RRs, while $CheA_3$ from the cytoplasmic cluster can only phosphotransfer to $CheB_2$ and $CheY_6$. The specificity was shown by co-crystal studies and mutagenesis to depend on a single methionine in the hydrophobic interaction domain, suggesting specificity between HK and RR proteins is defined by very few amino acids, and this allows the required phosphotransfer kinetics for a functional system (Bell *et al.*, 2010).

The experiments outlined above provided a set of data including *in vitro* phosphotransfer kinetics for all CheAs to all CheYs and CheBs and the dephosphorylation rates, and average copy numbers of the proteins in each cluster, in addition to cellular localization. The response kinetics for a step-down in attractant concentration (the dominant signal in *R. sphaeroides* as it has a motor bias close to 0.85), show very similar characteristics to that seen in *E. coli*, and over a wide range of background concentrations the response is dependent on the percentage change. These data were used to help suggest how the two chemosensory pathways might regulate the stopping activity of a single motor, and guide future experiments. A set of ordinary-differential equations (ODEs) representing the phosphotransfer and dephosphorylation kinetics of the polar and cytoplasmic clusters were derived and the effect on the phosphorylation state of the different RR proteins modelled (Fig. 3.9; Tindall *et al.*, 2010).

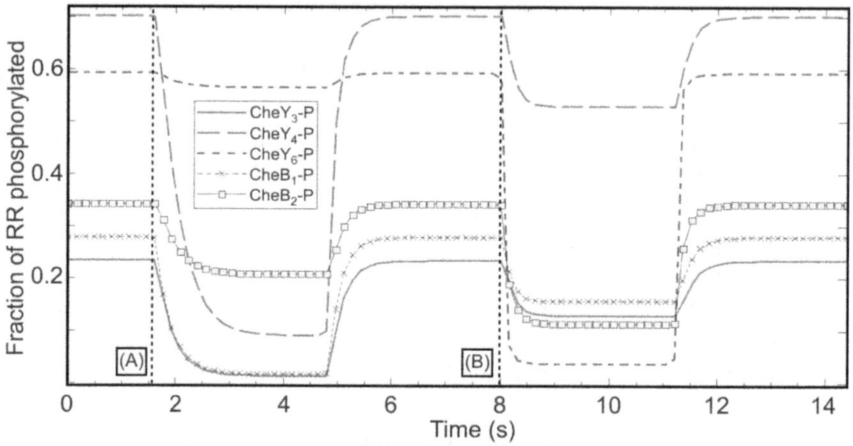

Figure 3.9 At time (A) $CheA_2$ at the polar cluster was inactivated, and as expected the phosphorylation state of all the RR proteins fell, while they were all increased when $CheA_2$ was active at about 5 s. Unexpectedly, when $CheA_3$ was inhibited at time (B), and then active at about 11 s, not only were $CheB_2$ and $CheY_6$ phosphorylation states altered, but also those of the other, polar RRs. This suggests that the cytoplasmic CheA, only able to directly phosphorylate two of the RR proteins can indirectly regulate the other, polar, RRs. Image adapted from Tindall *et al.* (2010).

The results were unexpected as they suggested a link between the two chemosensory clusters, with a change in the cytoplasmic CheA, only able *in vitro* to phosphotransfer to $CheY_6$ and $CheB_2$, able to alter the phosphotransfer rate of the polar RRs. The model also showed a faster dephosphorylation rate for $CheY_3$-P and $CheY_4$-P than measured *in vitro*.

In vitro experiments using the two RRs $CheB_2$-P and $CheY_6$-P showed, unexpectedly, that $CheB_2$-P could phosphotransfer to $CheA_2$, thus allowing a phosphorelay from $CheA_3$ to $CheB_2$ to $CheA_2$ and thus to $CheY_3$, $CheY_4$ and $CheB_1$. As $CheA_2$ can also phosphotransfer to $CheB_2$, and $CheB_2$ (unlike the *E. coli* CheB is diffuse in the cytoplasm) this provides a route for cytoplasmic, metabolic signals to tune the strength of the chemosensory response to extracellular compounds sensed via the cytoplasmic cluster. These experiments were all however done *in vitro*, and need to be tested *in vivo*. As it is currently impossible to identify phosphorylated RRs *in vivo* because of the kinetics, this will require future experiments, possible using FRET within *R. sphaeroides* to measure protein–protein interactions. This system may also be sensitive to the changes in relative expression levels of the different proteins, with altered levels of CheA and CheBs relative to each other potentially mopping up any phosphorelay activity. The possible effects of changing copy numbers on response kinetics and sensitivity require both computational modelling and experimental analysis.

The second prediction, that the CheYs dephosphorylate faster than measured *in vitro* suggests that the only identified phosphatase, a domain in the $CheA_3$ kinase which specifically dephosphorylates $CheY_6$-P, may act as a phosphate sink for the other RRs. This may explain why the HK on the cytoplasmic cluster is split across two proteins. $CheA_4$, the kinase, has a different receptor binding domain to $CheA_3$, with the Hpt histidine domain and the phosphatase domain, allowing the kinase and phosphatase activities to be independently regulated, again allowing tuning of the response with respect to the current metabolic state of the cell. Kinase–phosphatase pairs are found in many regulatory steps, including many TCSs, and are a common switching device in biology (Gao and Stock, 2009). However, it is rare to have the proteins arranged across two interconnected proteins. There is no evidence that the phosphatase activity of $CheA_3$ is in any way regulated by the phosphorylation state of the Hpt domain of $CheA_3$, phosphorylated by the kinase domain of $CheA_4$. It seems more likely that the different receptor binding domains tune the activity, but as yet we do not know whether the two CheAs bind to different Tlps or the same Tlps, but with different activities being stimulated. Interestingly if the receptor binding domains of $CheA_3$ and $CheA_4$ were swapped, they still localize to the cytoplasmic cluster, but do not support chemotaxis. This needs further investigation, as does the reason for there being two CheWs associated with the polar cluster. Informatic studies showed that across the sequenced microbial genomes there were a wide range of different chemoreceptors, categorized by the length of the coiled-coil cytoplasmic domain (Alexander and Zhulin, 2007). All the MCPs of *E. coli* belong to one family, but those of species with large numbers of chemoreceptor genes often belong to a number of families with different length cytoplasmic domains. Perhaps these require additional CheWs to coordinate the quaternary clusters and regulate a single CheA. Again this needs further data and predictive models based on interaction domains.

An alternative modelling route was also tried to suggest possible mechanisms for two pathways regulating one output. This used an approach common in control engineering where multiple models were created, each representing one of the possible system connectivities. Using the experimental data models were generated, each of which could represent those data. Where more than one parameter value is capable of fitting the wild-type data then multiple models can be produced, one containing each parameter that fit the data. This is particularly useful if there are model parameters that are difficult to measure experimentally and in addition will allow for parameter identification. Engineering tools can then be applied to determine whether there is an optimal input to the system, for example using a bode plot to determine if there is an optimal frequency which will result in different outputs between the models under test. When the determination of an optimal input do not result in successful model invalidation then changes to the initial conditions can be investigated. The advantage of using this approach is that multiple different possibilities can be tested *in silico* before performing additional experiments and the changes to the initial conditions can be selected, which result in the biggest difference between the models. Experimentally, changes in initial conditions can be realized by making changes in protein level. Robustness and sensitivity analysis can then be performed to ensure that the experiment will provide discrimination.

A number of models were produced that were consistent with the experimental output, and parameters changed experimentally and the resultant output measured against the models as part of an iterative cycle (Roberts *et al.*, 2010). This procedure eliminates models, as new experimental data did not fit specific model connectivities. The approach, however, cannot be used to validate models, as other unconsidered models with different methods of representing the interactions could still be valid. The approach does give insight into how the pathway may be wired, and reduces the number of complex experimental procedures needed to get closer to a realistic understanding of sensory signalling (Roberts *et al.*, 2010).

All the above experiments and models have dealt with the excitatory state of the chemosensory signal, and not with adaptation. *R. sphaeroides* responds by stopping in response to a step down in effector concentration and then adapts back to the prestimulus state. The response kinetic data suggests, as in *E. coli* there is robust and precise adaptation, however the percentage of cells responding and the background concentration over which the cells respond to a specific stimulus varies with growth conditions. The expression of the two pathways may also be under the control of different environmentally regulated sigma factors and that, along with the role of metabolism in generating a signal through the cytoplasmic cluster, adds another layer of complexity to the *R. sphaeroides* pathway, suggesting that the strength of the signals from the different clusters may vary with growth.

Chemotaxis requires $CheR_2$, $CheR_3$ and $CheB_1$ while deletion of $CheB_2$ reduces the response. This suggests that the major role of $CheB_2$ could be linking the two pathways and $CheB_1$ may be the major esterase. This requires much more experimental work. *In vivo* measurements showed that all the *R. sphaeroides* CheR and CheB proteins will complement *E. coli* deletions, when overexpressed, while *in vitro* analysis shows that $CheR_2$ and $CheR_3$ will methylate Tsr from *E. coli* suggesting that methylation, and therefore adaptation, is important for both the cytoplasmic Tlp and transmembrane MCP receptors. The precise role and mechanism of adaptation of the two clusters and how this generates response and adaptation kinetics similar to those seen in *E. coli* is much harder to test experimentally than in *E. coli*, and will need the help of predictive models to guide the experimental design. In *E. coli* a single chemoreceptor can be over expressed, and the adaptation sites mutated to cover the range of adaptation states, as described in the earlier section. The response and adaptation kinetics can be measured and the sensitivity and robustness fitted and tested in models. This is much more complex in *R. sphaeroides*. Bioinformatics has identified four glutamate/glutamine sites on each of the eight MCPs and two possible sites on each Tlp, the numbers of mutants required to create a full set of adaptation mutants in all combinations is experimentally prohibitive. In *E. coli* a single chemoreceptor can be overexpressed in a specific adaptation state and stimulated with an effector analogue. Unfortunately, deleting any Tlp results in the loss of the cytoplasmic cluster and chemotaxis (Wadhams *et al.*, 2005). In addition, none of the MCP receptors appears to respond to a single specific attractant. As has been identified in other species, deletion of one MCP results in a general reduction in the response to all attractants, while deletion of all and over expression of one, results in poor responses. This suggests the sensory regulation within the quaternary polar cluster may be different to *E. coli*, or the loss of the correct stoichiometry within or between clusters is critical. Adaptation therefore requires careful predictive modelling to guide the experiments to unravel the complexities of resetting the two systems. Current work centres on mutating all the potential adaptation sites on the Tlps to prevent adaptation and measuring any change in response kinetics and sensitivity. This will then be fitted into currently developed ODE models.

An alternative *in silico* method again uses control engineering to invalidate particular interactions networks by measuring the outputs as parameters in the connecting pathways as they are changed and when the copy numbers of the different CheBs are changed experimentally. Surprisingly this approach suggested that the feedback in the chemotaxis system might be arranged in an asymmetric fashion (Fig. 3.10). That is, $CheB_1$-P may only regulate adaptation of the polar MCPs while $CheB_2$-P may interact with both clusters, although the role in adaptation is unclear. This model suggests that *R. sphaeroides* may have a modular system that includes two feedback loops, one nested within the other. The nested loop is used to regulate a subprocess of the system whilst the 'external' negative feedback loop, as in *E. coli*, from the system output to the input is used to regulate the entire system (Hamadeh *et al.*, 2011). Again this will require rational experimental design to test the possibilities raised.

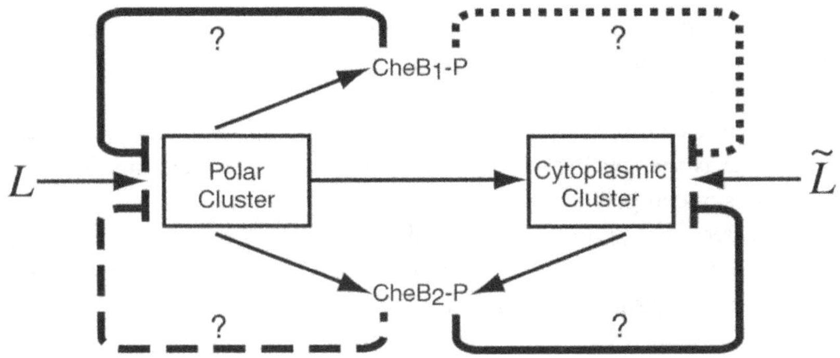

Figure 3.10 Control engineering view of nested, asymmetric feedback loops. The *R. sphaeroides* chemotaxis pathway with cascade control feedback offers the bacterium two feedback loops, one embedded within the other, to adapt and to reduce its sensitivity to parameter variations and noise, such as ligand input from the external and intracellular environments, L and $\tilde{L}$. This architecture of the system would be strongly sensitive to inputs to the cytoplasmic cluster, perhaps from the metabolic state of the cell. The feedback loops need to be experimentally tested, but explain current data. Image adapted from Hamadeh *et al.* (2011).

Discussion

In this chapter, we first reviewed our current understanding of *E. coli* chemotaxis using the MWC model as a guide. This model and its recent extensions (Clausznitzer *et al.*, 2010; Clinton *et al.*, 2010) can account extremely well for amplification of tiny stimuli and integration of multiple stimuli based on receptor cooperativity within small signalling complexes. Together with the Barkai and Leibler model for robust precise adaptation (Barkai and Leibler, 1997), and its newer, more elaborate versions (Endres and Wingreen, 2006; Clausznitzer *et al.*, 2010), the signalling pathway is relatively well understood. This provides a solid foundation for research into less established topics in bacterial chemotaxis as well as related bacterial and non-bacterial pathways. Here, in the second part of the chapter, we specifically reported on the application of similar experimental and modelling approaches to better understand *R. sphaeroides* chemotaxis. This bacterium reflects the architecture that may be present in many species with both membrane-embedded and cytoplasmic clusters, presumably to combine measurements of its external environment with its interior metabolic state, allowing it to live a versatile metabolic life style (Thompson *et al.*, 2006; Porter *et al.*, 2008). Indeed, chemotaxis in this bacterium is currently the subject of intense experimental study, using *in vitro* phosphorylation assays, chemotaxis-protein depletion and expression-from-plasmid assays, as well as *in vivo* motor response assays to better understand the subtleties of the kinetics (Thompson *et al.*, 2006; Porter *et al.*, 2008; Pilizota *et al.*, 2009). The future will show if an *in vivo* FRET system can be made work in *R. sphaeroides*, which was so beneficial in elucidating *E. coli* chemotaxis.

There are several open questions in *R. sphaeroides* chemotaxis which need to be addressed (Porter *et al.*, 2008, 2011):

1 What do the cytoplasmic and polar chemotaxis clusters sense? How is this sensory information integrated with signals from the membrane-bound chemotaxis cluster?

2 How does robust and precise adaptation work in this bacterium? $CheB_2$ may potentially be phosphorylated by both membrane-bound and cytoplasmic cluster-associated kinases (Roberts *et al.*, 2010; Hamadeh *et al.*, 2011).
3 Why does this bacterium have three CheYs to control the rotation of its single flagellar motor? What is the specific effect of each CheY on the motor, and do they bind to different sites or compete for one site?
4 Is the dedicated $CheY_6$ phosphatases a sink for the other CheYs?
5 How are the cytoplasmic receptors, CheWs and CheAs arranged, without the organizing framework of the cytoplasmic membrane?

Our theoretical understanding of *E. coli* chemotaxis, as well as information processing, signal amplification, signal integration, precise adaptation, and integral feedback control will probably prove beneficial in making progress. In this context, other organisms may be investigated as well, including *B. subtilis,* known for its three different adaptation systems (Rao *et al.*, 2004, 2008), and *Caulobacter crescentus,* which has attracted a large amount of interest due to its complex life style and development. The latter organism has already been used for detailed cryo-electron microscopy studies to elucidate the chemoreceptor arrangement in its clusters (Briegel *et al.*, 2008, 2009).

While *E. coli* chemotaxis seems exceptionally well understood, the thorough knowledge of its chemotaxis pathway, combined with the good understanding of other pathways in this bacterium, may provide a new wave of research in *E. coli.* Several obvious directions may be followed: expansion to other two-component systems, as well as cross-talk between other two-component systems and other pathways, including quorum sensing and metabolism. The Tsr receptor was recently found to allow chemotaxis to the interspecies quorum-sensing signal AI-2 via the periplasmic LsrB AI-2-binding protein (Hegde *et al.*, 2010). The cross-talk with metabolism via the phosphotransferase system (PTS), fumarate and acetyl phosphate is well documented. Carbohydrates are either sensed via periplasmic binding proteins and receptors independent of uptake, or are sensed by PTS while taken up (Lux *et al.*, 1999; Szurmant and Ordal, 2004). The latter may serve as a way to sense glucose if periplasmic binding proteins are saturated by an inferior ligand (Neumann *et al.*, 2010). In PTS, carbohydrates are phosphorylated and trapped upon entry via substrate phosphoenolpyruvate (PEP). There are several carbohydrate-specific membrane-embedded enzymes for uptake, and common cytosolic enzyme I is believed to interact with CheA to regulated chemotaxis. Interestingly *R. sphaeroides* only shows a strong response to one sugar, fructose, which is transported by an unusual PTS system, the only PTS shown in *E. coli* not to integrate with the chemosensory system. In *R. sphaeroides* the response to fructose seems to directly link to metabolism and therefore probably the cytoplasmic cluster. Acetyl phosphate levels in *E. coli* directly reflect the metabolic state of the cell and may tune the strength of any chemosensory response.

In addition to cross-talk with metabolism, there are also more general questions to be followed up on. For example, how is chemotaxis related to virulence, symbiotic behaviour, and complex cellular communities (Terry *et al.*, 2005; Boehm *et al.*, 2010)? Finally, is it possible to use bacterial chemotaxis in novel applications in synthetic biology or the pharmaceutical industry? For instance, the periplasmic ligand-binding domains of chemoreceptors may be redesigned to allow the use of bacteria as bio-sensors (Kiessling *et al.*, 2006). Furthermore, bacteria have been considered as vehicles for the invasion and destruction of tumours

(Anderson *et al.*, 2006). Studies on *E. coli* chemotaxis are on the verge of reaping the practical benefits of decades of basic research endeavour.

Acknowledgements

R.G.E. acknowledges funding from the Biotechnology and Biological Sciences Research Council grant BB/G000131/1 and from the Centre for Integrative Systems Biology at Imperial College (CISBIC). Research in the Armitage lab is funded by BBSRC.

References

Alexander, R.P., and Zhulin, I.B. (2007). Evolutionary genomics reveals conserved structural determinants of signaling and adaptation in microbial chemoreceptors. Proc. Natl. Acad. Sci. U.S.A. *104*, 2885–2890.

Alon, U., Surette, M.G., Barkai, N., and Leibler, S. (1999). Robustness in bacterial chemotaxis. Nature *397*, 168–171.

Anderson, J.C., Clarke, E.J., Arkin, A.P., and Voigt, C.A. (2006). Environmentally controlled invasion of cancer cells by engineered bacteria. J. Mol. Biol. *355*, 619–627.

Aquino, G., Clausznitzer, D., Tollis, S., and Endres, R.G. (2011). Optimal receptor-cluster size determined by intrinsic and extrinsic noise. Phys. Rev. E *83*, 021914.

Asakura, S., and Honda, H. (1984). Two-state model for bacterial chemoreceptor proteins: the role of multiple methylation. J. Mol. Biol. *176*, 349–367.

Bai, F., Branch, R.W., Nicolau, D.V., Jr., Pilizota, T., Steel, B.C., Maini, P.K., and Berry, R.M. (2010). Conformational spread as a mechanism for cooperativity in the bacterial flagellar switch. Science *327*, 685–689.

Bansal, T., Alaniz, R.C., Wood, T.K., and Jayaraman, A. (2010). The bacterial signal indole increases epithelial-cell tight-junction resistance and attenuates indicators of inflammation. Proc. Natl. Acad. Sci. U.S.A. *107*, 228–233.

Barkai, N., and Leibler, S. (1997). Robustness in simple biochemical networks. Nature *387*, 913–917.

Bell, C.H., Porter, S.L., Strawson, A., Stuart, D.I., and Armitage, J.P. (2010). Using structural information to change the phosphotransfer specificity of a two-component chemotaxis signalling complex. PLoS Biol. *8*, e1000306.

Berg, H.C. (2000). Motile behavior of bacteria. Phys. Today *53*, 24–29.

Berg, H.C., and Brown, D.A. (1972). Chemotaxis in *Escherichia coli* analysed by three-dimensional tracking. Nature *239*, 500–504.

Boehm, A., Kaiser, M., Li, H., Spangler, C., Kasper, C.A., Ackermann, M., Kaever, V., Sourjik, V., Roth, V., and Jenal, U. (2010). Second messenger-mediated adjustment of bacterial swimming velocity. Cell *141*, 107–116.

Bray, D., Levin, M. D., and Morton-Firth, C. J. (1998). Receptor clustering as a mechanism to control sensitivity. Nature *393*, 85–88.

Briegel, A., Ding, H.J., Li, Z., Werner, J., Gitai, Z., Dias, D.P., Jensen, R.B., and Jensen, G.J. (2008). Location and architecture of the *Caulobacter crescentus* chemoreceptor array. Mol. Microbiol. *69*, 30–41.

Briegel, A., Ortega, D.R., Tocheva, E.I., Wuichet, K., Li, Z., Chen, S., Müller, A., Iancu, C.V., Murphy, G.E., Dobro, M.J., Zhulin, I.B., *et al.* (2009). Universal architecture of bacterial chemoreceptor arrays. Proc. Natl. Acad. Sci. U.S.A. *106*, 17181–17186.

Clausznitzer, D., Oleksiuk, O., Løvdok, L., Sourjik, V., and Endres, R.G. (2010). Chemotactic response and adaptation dynamics in *Escherichia coli*. PLoS Comput. Biol. *6*, e1000784.

Cluzel, P., Surette, M., and Leibler, S. (2000). An ultrasensitive bacterial motor revealed by monitoring signaling proteins in single cells. Science *287*, 1652–1655.

Duke T.A., and Bray, D. (1999). Heightened sensitivity of a lattice of membrane receptors. Proc. Natl. Acad. Sci. U.S.A. *96*, 10104–10108.

Dusenbery, D.B. (1998). Spatial sensing of stimulus gradients can be superior to temporal sensing for free-swimming bacteria. Biophys. J. *74*, 2272–2277.

Emonet, T., and Cluzel, P. (2008). Relationship between cellular response and behavioral variability in bacterial chemotaxis. Proc. Natl. Acad. Sci. U.S.A. *105*, 3304–3309.

Endres, R.G. (2009). Polar chemoreceptor clustering by coupled trimers of dimers. Biophys. J. *96*, 453–463.

Gao R., and Stock, A.M. (2009). Biological insights from structures of two-component proteins, Annu. Rev. Microbiol. *63,* 133–154.

Gauger, E.J., Leatham, M.P., Mercado-Lubo, R., Laux, D.C., Conway, T., and Cohen, P.S. (2007). Role of motility and the flhDC operon in *Escherichia coli* MG1655 colonization of the mouse intestine. Infect. Immun. *75,* 3315–3324.

Gestwicki, J.E., Lamanna, A.C., Harshey, R.M., McCarter, L.L., and Kiessling L.L. (2000). Evolutionary conservation of methyl-accepting chemotaxis protein location in bacteria and archaea. J. Bacteriol. *182,* 6499–6502

Greenfield, D., McEvoy, A.L., Shroff, H., Crooks, G.E. Wingreen, N.S., Betzig, E., and Liphardt, J. (2009). Self-organization of the *Escherichia coli* chemotaxis network imaged with super-resolution light microscopy. PLoS Biol. *7,* e1000137.

Greer-Phillips, S.E., Stephens, B.B., and Alexandre, G. (2004). An energy taxis transducer promotes root colonization by Azospirillum brasilense. J. Bacteriol. *186,* 6595–6604.

Hamadeh, A., Roberts, M.A.J., August, E., McSharry, P.E., Maini, P.K., Armitage, J.P., and Papachristodoulou, A. (2011). Feedback control architecture and the bacterial chemotaxis network. PLoS Comp. Biol. *7,* e1001130.

Hansen, C.H., Sourjik, V., and Wingreen, N.S. (2010). A dynamic-signaling-team model for chemotaxis receptors in *Escherichia coli.* Proc. Natl. Acad. Sci. U.S.A. *107,* 17170–17175.

Hegde, M., Englert, D.L., Schrock, S., Cohn, W.B., Vogt, C., Wood, T.K., Manson, M.D., and Jayaraman, A. (2011). Chemotaxis to the quorum-sensing signal ai-2 requires the Tsr chemoreceptor and the periplasmic LsrB AI-2-binding protein. J. Bacteriol. *193,* 768–773.

Jeziore-Sassoon, Y., Hamblin, P.A., Bootle-Wilbraham, C.A., Poole, P.S., and Armitage, J.P. (1998). Metabolism is required for chemotaxis to sugars in Rhodobacter sphaeroides. Microbiology *144,* 229–239.

Kalinin, Y., Neumann, S., Sourjik, V., and Wu, M. (2010). Responses of *Escherichia coli* bacteria to two opposing chemoattractant gradients depend on the chemoreceptor ratio. J. Bacteriol. *192,* 1796–1800.

Keymer, J. E., Endres, R. G., Skoge, M., Meir, Y., and Wingreen, N. S. (2006). Chemosensing in *Escherichia coli*: two regimes of two-state receptors. Proc. Natl. Acad. Sci. U.S.A. *103,* 1786–1791.

Kiessling, L.L., Gestwicki, J.E., and Strong, L.E. (2006). Synthetic multivalent ligands as probes of signal transduction. Angew. Chem. Int. Ed. Engl. *45,* 2348–2368.

Kirby, J.R. (2010). Synthetic biology: Designer bacteria degrades toxin. Nat. Chem. Biol. *6,* 398–399.

Korobkova, E., Emonet, T., Vilar, J.M., Shimizu, T.S., and Cluzel, P. (2004). From molecular noise to behavioural variability in a single bacterium. Nature *428,* 574–578.

Laub, M.T., and Goulian, M. (2007). Specificity in two-component signal transduction pathways. Ann. Rev. Genet. *41,* 121–145.

Lai, W.C., Barnakova, L.A., Barnakov, A.N., and Hazelbauer, G.L. (2006a). Similarities and differences in interactions of the activity-enhancing chemoreceptor pentapeptide with the two enzymes of adaptational modification. J. Bacteriol. *188,* 5646–5649.

Lai, W.C., Beel, B.D., and Hazelbauer, G.L. (2006b). Adaptational modification and ligand occupancy have opposite effects on positioning of the transmembrane signalling helix of a chemoreceptor. Mol. Microbiol. *61,* 1081–1090.

Li, M., and Hazelbauer, G.L. (2004). Cellular stoichiometry of the components of the chemotaxis signaling complex. J. Bacteriol. *186,* 3687–3694.

Li, M., and Hazelbauer, G. L. (2005). Adaptational assistance in clusters of bacterial chemoreceptors. Mol. Microbiol. *56,* 1617–1626.

Li, M., and Hazelbauer, G.L. (2006). The carboxyl-terminal linker is important for chemoreceptor function. Mol. Microbiol. *60,* 469–479.

Lux, R., Munasinghe, V.R., Castellano, F., Lengeler, J.W., Corrie, J.E., and Khan, S. (1999). Elucidation of a PTS-carbohydrate chemotactic signal pathway in *Escherichia coli* using a time-resolved behavioral assay. Mol. Biol. Cell. *10,* 1133–1146.

Mackenzie, C., Choudhary, M., Larimer, F.W., Predki, P.F., Stilwagen, S., Armitage, J.P., Barber, R.D., Donohue, T.J., Hosler, J.P., Newman, J.E., *et al.* (2001). The home stretch, a first analysis of the nearly completed genome of *Rhodobacter sphaeroides* 2.4.1. Photosynth. Res. *70,* 19–41.

Maddock, J. R., and Shapiro, L. (1993). Polar localization of the chemoreceptor complex in the *Escherichia coli* cell. Science *259,* 1717–1723.

Mao, H., Cremer, P.S., and Manson, M.D. (2003). A sensitive, versatile microfluidic assay for bacterial chemotaxis, Proc. Natl. Acad. Sci. U.S.A. *100,* 5449–5454.

Meir, Y., Jakovljevic, V., Oleksiuk, O., Sourjik, V., and Wingreen, N.S. (2010). Precision and kinetics of adaptation in bacterial chemotaxis. Biophys. J. *99*, 2766–2774.

Mesibov, R., Ordal, G.W., and Adler, J. (1973). The range of attractant concentrations for bacterial chemotaxis and the threshold and size of response over this range. Weber law and related phenomena. J. Gen. Physiol. *62*, 203–223.

Miller, L.D., Yost, C.K., Hynes, M.F., and Alexandre, G. (2007). The major chemotaxis gene cluster of *Rhizobium leguminosarum* bv. *viciae* is essential for competitive nodulation. Mol. Microbiol. *63*, 348–362.

Monod, J., Wyman, J., and Changeux, J. P. (1965). On the nature of allosteric transitions: a plausible model. J. Mol. Biol. *12*, 88–118.

Neumann, S., Hansen, C.H., Wingreen, N.S., and Sourjik, V. (2010). Differences in signalling by directly and indirectly binding ligands in bacterial chemotaxis. EMBO J. *29*, 3484–3495.

Pilizota, T., Brown, M.T., Leake, M.C., Branch, R.W., Berry, R.M., and Armitage, J.P. (2009). A molecular brake, not a clutch, stops the *Rhodobacter sphaeroides* flagellar motor. Proc. Natl. Acad. Sci. U.S.A. *106*, 11582–11587.

Porter, S.L., and Armitage, J.P. (2002). Phosphotransfer in *Rhodobacter sphaeroides* chemotaxis. J. Mol. Biol. *324*, 35–45.

Porter, S.L., Wadhams, G.H., Martin, A.C., Lancaster, D.E., Byles, E., and Armitage, J.P. (2006). The CheYs of *Rhodobacter sphaeroides*. J. Biol. Chem. *281*, 32694–32704.

Porter, S.L., Roberts, M.A., Manning, C.S., and Armitage, J.P. (2008a). A bifunctional kinase-phosphatase in bacterial chemotaxis. Proc. Natl. Acad. Sci. U.S.A. *105*, 18531–18536.

Porter, S.L., Wadhams, G.H., and Armitage, J.P. (2008b). *Rhodobacter sphaeroides*: complexity in chemotactic signalling. Trends Microbiol. *16*, 251–260.

Porter, S.L., Wadhams, G.H., and Armitage, J.P. (2011). Signal processing in complex chemotaxis pathways. Nature Rev. Microbiol. *9*, 153–165.

Postma, P.W., Lengeler, J.W., and Jacobson, G.R. (1993). Phosphoenolpyruvate: carbohydrate phosphotransferase systems of bacteria. Microbiol. Rev. *57*, 543–594.

Rawls, J.F., Mahowald, M.A., Goodman, A.L., Trent, C.M., and Gordon, J.I. (2007). *In vivo* imaging and genetic analysis link bacterial motility and symbiosis in the zebrafish gut. Proc. Natl. Acad. Sci. U.S.A. *104*, 7622–7627.

Roberts, M.A., Papachristodoulou, A., and Armitage, J.P. (2010). Adaptation and control circuits in bacterial chemotaxis. Biochem. Soc. Trans. *38*, 1265–1269.

Salman, H., and Libchaber, A. (2007). A concentration-dependent switch in the bacterial response to temperature. Nat. Cell. Biol. *9*, 1098–1100.

Scott, K.A., Porter, S.L., Bagg, E.A.L., Hamer, R., Hill, J.L., Wilkinson, D., and Armitage, J.P. (2010). Specificity of localization and phosphotransfer in the CheA proteins of *Rhodobacter sphaeroides*. Mol. Microbiol. *76*, 318–330.

Singh, A.H., Wolf, D.M., Wang, P., and Arkin, A.P. (2008). Modularity of stress response evolution. Proc. Natl. Acad. Sci. U.S.A. *105*, 7500–7505.

Skoge, S., Endres, R. G., and Wingreen, N. S. (2006). Receptor–receptor coupling in bacterial chemotaxis: evidence for strongly coupled clusters. Biophys. J. *90*, 4317–4326.

Sourjik, V. (2004). Receptor clustering and signal processing in *E. coli* chemotaxis. Trends Microbiol. *12*, 569–576.

Sourjik, V., and Berg, H. C. (2002). Receptor sensitivity in bacterial chemotaxis. Proc. Natl. Acad. Sci. U.S.A. *99*, 123–127.

Sourjik, V., and Berg, H. C. (2004). Functional interactions between receptors in bacterial chemotaxis. Nature *428*, 437–441.

Szurmant, H., and Ordal, G.W. (2004). Diversity in chemotaxis mechanisms among the bacteria and archaea. Microbiol. Mol. Biol. Rev. *68*, 301–319.

Terry, K., Williams, S.M., Connolly, L., and Ottemann, K.M. (2005). Chemotaxis plays multiple roles during *Helicobacter pylori* animal infection. Infect. Immun. 73, 803–811.

Thiem, S., Kentner, D., and Sourjik, V. (2007). Positioning of chemosensory clusters in *E. coli* and its relation to cell division. EMBO J. *26*, 1615–1623.

Tindall, M.J., Porter, S.L., Maini, P.K., and Armitage, J.P. (2010). Modeling chemotaxis reveals the role of reversed phosphotransfer and a bi-functional kinase-phosphatase. PLoS Comp. Biol. *6*, e1000896.

Thompson, S.R., Wadhams, G.H., and Armitage, J.P. (2006). The positioning of cytoplasmic protein clusters in bacteria. Proc. Natl. Acad. Sci. U.S.A. *103*, 8209–8214.

Toews, M.L., Goy, M.F., Springer, M.S., and Adler, J. (1979). Attractants and repellents control demethylation of methylated chemotaxis proteins in *Escherichia coli*. Proc. Natl. Acad. Sci. U.S.A. *76*, 5544–5548.

Wadhams ,G.H., and Armitage, J.P. (2004). Making sense of it all: bacterial chemotaxis. Nat. Rev. Mol. Cell. Biol. *5*, 1024–1037.

Wadhams, G.H., Warren, A.V., Martin, A.C., and Armitage, J.P. (2003). Targeting of two signal transduction pathways to different regions of the bacterial cell. Mol. Microbiol. *50*, 763–770.

Wadhams, G.H., Martin, A.C., and Armitage, J.P. (2005). Factors determining formation of polar and cytoplasmic chemosensory clusters in *Rhodobacter sphaeroides*. Mol. Microbiol. *58*, 895–902.

Williams, S.M., Chen, Y.T., Andermann, T.M., Carter, J.E., McGee, D.J., and Ottemann, K.M. (2007). Helicobacter pylori chemotaxis modulates inflammation and bacterium-gastric epithelium interactions in infected mice. Infect. Immun. *75*, 3747–3757.

Wuichet, K., and Zhulin, I.B. (2010). Origins and diversification of a complex signal transduction system in prokaryotes. Sci. Signal. *3*, ra50.

Wuichet, K., Cantwell, B.J., and Zhulin, I.B. (2010). Evolution and phyletic distribution of two-component signal transduction systems. Curr. Opin. Microbiol. *13*, 219–225.

Yi, T-.M., Huang, Y., Simon, M. I., and Doyle, J. (2000). Robust perfect adaptation in bacterial chemotaxis through integral feedback control. Proc. Natl. Acad. Sci. U.S.A. *97*, 4649–4653.

Systems Biology of Infection: The Pathogen Perspective

4

Dirk Bumann

Abstract

Microbial infections still cause around one-quarter of all deaths globally, despite the advances that have been made in the treatment of infectious disease. The increasing occurrence of drug resistant pathogens, both old and new, coupled with an increasingly mobile human population has created many novel opportunities for potential pathogens to meet new human hosts. All of this requires new prevention and control strategies but progress has been slow, despite recent technological advances and increased investment. The rapid increase in data proved difficult to translate into practical applications for human health care, and new approaches and analyses are needed to make the most of new opportunities. One of these is the use of the new tools of systems biology and this chapter will review the application of these to microbial pathogens.

The need for systems biology of infection

Despite advances in the treatment of infectious disease, pathogenic bacteria represent one of the most important threats to human health worldwide. The inexorable rise in antibiotic resistance, (re-)emerging pathogens, climate change and increasing human mobility contribute to new scenarios posed by infectious agents, that require new control strategies (Arias and Murray, 2009). Unfortunately, progress in both vaccine development and antimicrobial chemotherapy has been slow despite major investments from industry, public agencies and charitable funds (Payne *et al.*, 2007). This is even more surprising since infectious disease research has flourished in the past 20 years yielding detailed insights into aspects ranging from molecular disease mechanisms to epidemiology. However, this burgeoning new knowledge has remained difficult to translate into practical applications for human health care.

These difficulties represent a motivation to explore novel research strategies. Most current research focuses on the role of specific pathogen or host factors. While this reductionist approach has been extremely successful in elucidating a large number of important virulence and immune mechanisms, it does not take into account the complex interplay of hundreds of pathogen host factors that is key for a successful pathogen. In addition to reductionist approaches, a systems-level analysis is thus necessary to understand infectious diseases and to develop rational and novel control strategies.

As an example, type III or type IV secretion of pathogen effectors into the host cell cytosol is essential for virulence of diverse pathogens (Galan, 2009). Many of these secretion systems translocate dozens to hundreds of effectors, but the specific role of individual effectors is often difficult to identify because of extensive functional overlap. Only systematic analysis of defined combinations of effectors can unravel individual contributions and collective functions.

Another example is the pathogen stress responses that are often critical for virulence. However, stress stimuli often induce hundreds of differentially expressed genes most of which make only small individual contributions to virulence. It is thus a combination of interacting genes that is critical for stress responses that can protect pathogens against host defence mechanisms. Again, a purely reductionist approach is unlikely to yield a true understanding of the infection.

In terms of application, antimicrobial chemotherapy has been highly successful in the past and is likely to remain an important approach to combat infection. Unfortunately, the number of suitable individual pathogen targets appears to be rather limited (Becker *et al.*, 2006) suggesting that combination approaches with parallel inhibition of multiple targets might be needed. Because of the vast number of possible target combinations, development of such strategies is challenging and requires a systems-level approach.

Current status of systems biology of infection

Model systems to study infection biology

The ultimate goal of systems biology of infection is a systems-level understanding of the interaction of two different organisms with vastly different levels of complexity. It is thus necessary to investigate conditions in infected host animals/humans. However, ethical and technical limitations prevent detailed analysis of many important aspects under these relevant *in vivo* conditions. There is, therefore, a need for simpler model systems ranging from pure *in vitro* cultures over cell culture infections to small animal infection models. It is important to consider opportunities and limitations of such model systems.

Most infection biology genetics is undertaken with specific, well-characterized pathogen isolates. Advantages of using standard strains include availability of optimized experimental methods and often extensive literature facilitating interpretation of new results. On the other hand, individual common laboratory strains might not be representative of the species particularly since repeated *in vitro* passage often leads to loss of important virulence determinants.

In vitro cultures provide full control of experimental conditions and are often necessary to elucidate individual gene functions in detail. On the other hand, *in vitro* cultures can only partially mimic relevant *in vivo* conditions, and differential pathogen adaptation can result in major physiological differences that affect gene function.

Cell culture infection models include both pathogen and host cells and often successfully recapitulate key aspects of disease such as pathogen/host cell killing, intracellular pathogen growth and pathogen antigen presentation by host cells. Cell culture models enable high-resolution, time-lapse microscopy, in-depth biochemical analysis, and high-throughput RNA interference screening of host gene function. On the other hand, most cell culture infection models employ immortal cancer cell lines because these are available in unlimited amounts. However, cancer cells differ in many important aspects such as chromosome number, metabolism, and vesicular trafficking compared with primary cells that are more relevant *in vivo*. Moreover, tissue context can modulate host cell properties that might not be fully reflected in simplified cell culture models.

Mouse infection models enable analysis of host–pathogen interactions in complex tissues. The availability of a vast number of genetically modified mouse strains provides opportunities to systematically investigate the impact of host genes on pathogen properties and the course of infection. On the other hand, besides ethical concerns with animal experimentation, mouse infections might not fully represent conditions in more relevant (human) hosts, especially in cases where the mouse is not the natural host for the pathogen of interest. Moreover, most genetically modified mice have a specific strain background (C57BL/6) that carries a defective allele of *Slc11a1* (*NRAMP*) which makes these mice highly susceptible especially to intracellular pathogens (Bellamy, 1999).

Human studies are in many cases most relevant to understand infectious diseases and to critically assess potential novel strategies for disease control. However, ethical aspects including confidentiality of personal data and experimental restriction limit investigational strategies. Moreover, important variability in genotype and lifestyle can confound the data.

Data sources

Systems biology of infection depends on the integration of complementary large-scale data sets. Such data sets are increasingly available for multiple pathogens owing to ever increasing data acquisition capacity.

Genomics

Complete genome sequences are now available for almost all important pathogens. Recently, deep-sequencing techniques have revolutionized genome sequencing enabling the rapid determination of genetic data for hundreds of isolates within a clonal lineage. These data are invaluable for global transmission tracking and to ascertain the evolution of lineages particularly relating to antibiotic resistance (Croucher *et al.*, 2011) . In addition to the identification of new genetic markers for specific strains, genetic differences between isolates of known provenance could also contribute to different disease outcomes. Limited comparisons of a few strains have often failed to reveal differential virulence determinants but the emerging massive data sets for hundreds of strains might provide the statistical power that is necessary to identify genome-wide association patterns for virulence mechanisms and disease outcome similar to the breakthrough in identifying genotype–phenotype associations in human genetics (Wang, K., *et al.*, 2010a).

Gene expression profiling

In a particular host environment, pathogens usually express only a subset of the genes. This specific adaptation often has a major impact on pathogen properties and disease outcome. Various approaches have been developed to characterize gene expression at the transcript and protein level. In particular, a large number of transcriptome studies have revealed the extensive plasticity of pathogen gene expression in model systems from *in vitro* cultures to infected host tissues. Novel RNA-seq methods reveal pathogen transcriptomics at unprecedented resolution and sensitivity (Sharma *et al.*, 2010).

Transcript levels have been most extensively studied, but gene function usually depends on a protein as the final gene product. Proteomics has revealed large-scale data on pathogen protein content in various model systems. Recent developments have greatly increased sensitivity and detection speed resulting in increased coverage. Moreover, various methods employing isotopically labelled internal standards – but also label-free methods – enable relative and even absolute quantification at different levels of accuracy (Malmstrom *et al.*, 2009). Many proteins are modulated in their activity through post-translational modifications. Although this important issue has not yet been incorporated in most pathogen proteomics studies, suitable techniques for large-scale analysis become increasingly available to researchers (Macek *et al.*, 2008; Wang, Q., *et al.*, 2010).

Many gene expression studies include a pure *in vitro* culture as a reference for comparison. Such comparisons highlight specific pathogen adaptations to host cells and might thus enrich for relevant virulence mechanisms. However, in many cases the choice of *in vitro* conditions is somewhat arbitrary and severely influences comparison results. This might explain why sometimes '*in vivo* inducibility' of genes poorly correlates with their importance in virulence. In fact, genes expressed under some *in vitro* conditions can still serve important functions during infection and might actually represent integral members of virulence networks. For a system-level analysis of pathogen properties during infection it thus might be beneficial to include any gene that is expressed during host contact regardless of its expression in some arbitrary *in vitro* condition.

Almost all pathogen gene expression data are derived from large populations and thus represent average expression patterns. However, it is clear that individual pathogen cells can vary dramatically in their expression patterns as a result of internal stochastic variations or differential microenvironments (Eldar and Elowitz, 2010). In extreme cases, average expression patterns might substantially differ from what is present in any individual cell. Such heterogeneity can have a strong impact on disease course and treatment efficacy and therefore merits detailed analysis. Emerging methods to analyse heterogeneous subpopulations could include single-cell transcriptomics and proteomics of separated subpopulations (Khan *et al.*, 2005; Wang and Bodovitz, 2010).

Metabolomics

Metabolism is a prerequisite for pathogen growth and its regulation is finely tuned and strongly coordinated with pathogen virulence and gene expression. Metabolomics studies on pathogens analysed nutrient uptake and catabolism in both *in vitro* cultures and cell culture infection models using mass spectrometry and/or NMR spectroscopy. Currently, most analysis is limited to the central carbon metabolism and a small number of nutrients but new technologies greatly expand metabolite identification and tracking of isotopic labels through the entire metabolic network.

Mutant phenotypes

To assess the functional relevance of individual genes and gene networks for pathogen virulence, one needs to determine respective mutant phenotypes in infection models. In addition to conventional phenotyping of individual mutants, new high throughput approaches yield large-scale functional data sets. Large pools of mutants with defined gene deletions or transposon insertions can be administered to cell culture or animal infection models. Mutants with decreased fitness are identified based on their under-representation during infection using hybridization microarrays or deep-sequencing [STM, signature-tagged mutagenesis (Mazurkiewicz *et al.*, 2006; Gawronski *et al.*, 2009); TraDIS (Langridge *et al.*, 2009)]. The resulting data offer highly valuable large-scale insights in relevant virulence mechanisms.

Interestingly, mutant phenotyping of diverse pathogens has usually revealed a large majority of genes with no detectable impact on virulence. It is important to note that phenotypes that escape detection could be the result of true dispensability, functional redundancy among different genes that collectively mediate important functions, or simply technical issues that limit phenotype detection sensitivity. Unfortunately, detection limits are usually not quantified but in our own experience with a well-characterized mouse typhoid fever model, growth rate differences below a few per cent are difficult to detect (Becker *et al.*, 2006). This indicates that many minor contributions might escape detection. A group of several such minor virulence factors could together account for a substantial part of pathogenicity and this would have obvious relevance for any system-level understanding. It is thus mandatory to optimize sensitivity as much as possible.

Even in cases where mutations have clear virulence phenotypes, the underlying molecular mechanisms are often rather unclear. One basic problem is that phenotyping usually reflects just net growth. Any defect could be the result of decreased growth and/or enhanced pathogen killing. Partition of non-replicating plasmids or stable fluorescent proteins (Helaine *et al.*, 2010) can be used to track pathogen cell division and thereby to distinguish growth from killing. Another problem for phenotype interpretations is the lack of suitable readouts to more specifically characterize pathogen physiological dysfunction. As an example, it is often unclear how oxidative stress kills pathogens (i.e. which molecular damage of what targets is relevant and why does it kill the pathogen). A third problem for phenotype interpretation is that molecular mechanisms are typically elucidated in *in vitro* cultures under conditions that can be dramatically different from more relevant *in vivo* conditions. As an example, pathogen mutants with increased sensitivity to oxidative stress *in vitro* might show no phenotype in host environments with prominent oxidative stress. One potential explanation could be an *in vivo* upregulation of detoxifying systems that protect the pathogen even if one other defence system is dysfunctional. Better tools to analyse relevant molecular mechanisms leading to growth inhibition or increased pathogen killing are clearly needed as a precondition for any system-level understanding.

Data integration and analysis

Technological development yields increasingly accurate and comprehensive large-scale data sets on various aspects of pathogen biology during infection. As a first step to exploit these data for obtaining potential biological insights, and as a prerequisite to enable the full power of system-level analysis, such data sets should be integrated into a single database that contains genome comparisons, genome annotations, functional data with direct links to literature references, gene expression data, and visualization/analysis tools for modules, pathways, subnetworks and large networks. EcoCyc (Keseler *et al.*, 2011), Reactome (Croft *et al.*, 2011), and String (Szklarczyk *et al.*, 2011) are databases that include at least some of these aspects, but to date such a database does not exist for any pathogen/infectious disease. Challenges include incorporation of different data formats, establishment of a powerful database structure with extensive cross-referencing, and the large efforts required for manual curation.

Large-scale data sets also pose substantial challenges for analysis. A simple approach that can lead to striking results consists of cherry-picking a few genes with suspicious data such as the strongest virulence phenotypes or the most dramatic expression differences. Follow-up can then be focused on these individual factors. While such a reductionist approach is often used with great success, it falls short of fully exploiting network data. More integrated approaches are still in their infancy and are not yet widely used to analyse pathogen biology during infection.

As one alternative, clustering of genes/proteins/metabolites with similar properties can reveal modules of interacting factors (Bumann, 2010). Using a 'guilt-by-association' approach, a few well-characterized factors within such modules can help to understand the function of other previously uncharacterized factors. Unexpected clustering of seemingly unrelated factors can also reveal modular relations that structure broad physiological properties. However, experimental noise and incomplete/wrong functional information can result in artificial clusters and misleading interpretations. Appropriate validation with indicative mutations/mutation combinations is therefore required to confirm new modules and their potential relevance in virulence.

A more advanced type of analysis is possible for cellular activities with already well-characterized underlying network structures such as proteins interaction networks, signalling networks, or metabolism. Large-scale experimental data sets can then be superimposed on such structures to visualize prominently represented subnetworks, modules, and pathways as well as their interconnection. This analysis can efficiently guide functional validation by highlighting both central core factors as well as key connectors of distinct modules within the virulence associated network structure (Yoon *et al.*, 2011).

The most advanced approach would be integration of large-scale experimental data sets with genome scale *in silico* models. This is currently feasible only for the metabolism subsystem (Oberhardt *et al.*, 2009). Metabolites, metabolic reactions and enzymes that catalyse them, as well as pathways and super pathways have been studied in detail over many decades providing a suitable basis for genome-scale in silico reconstructions of metabolic networks for diverse pathogens. Computer simulations using these network reconstructions enable comprehensive phenotype predictions. Interestingly, the vast majority of these predictions agree with experimental data for *in vitro* cultures, and the few discrepancies are actually useful for highlighting knowledge gaps and areas for further experimentation. On the other hand, modelling of pathogen metabolism in host tissues remains challenging, particularly because of poorly defined microenvironments.

Challenges and outlook

System-level analysis of infectious diseases represents a new approach that is likely to offer a more appropriate understanding of complex interacting networks of pathogen and host factors. However this approach is still in its infancy and major experimental and theoretical challenges still need to be overcome. In particular, we need better diagnostic tools to understand the molecular basis of virulence phenotypes, accessible databases that combine diverse data sets, and well-established pipelines for integrative data analysis. Finally, it will be challenging to communicate complex findings on large-scale networks, and potentially even more challenging to extract from this knowledge a few crucial pathogen factors that could be inhibited alone or in combination to control infectious diseases.

References

Arias, C.A., and Murray, B.E. (2009). Antibiotic-resistant bugs in the 21st century – a clinical super-challenge. N. Engl. J. Med. *360*, 439–443.

Becker, D., Selbach, M., Rollenhagen, C., Ballmaier, M., Meyer, T.F., Mann, M., and Bumann, D. (2006). Robust *Salmonella* metabolism limits possibilities for new antimicrobials. Nature *440*, 303–307.
Bellamy, R. (1999). The natural resistance-associated macrophage protein and susceptibility to intracellular pathogens. Microbes Infect. *1*, 23–27.
Bumann, D. (2010). Pathogen proteomes during infection: a basis for infection research and novel control strategies. J. Proteomics 73, 2267–2276.
Croft, D., O'Kelly, G., Wu, G., Haw, R., Gillespie, M., Matthews, L., Caudy, M., Garapati, P., Gopinath, G., Jassal, B., *et al.* (2011). Reactome: a database of reactions, pathways and biological processes. Nucleic Acids Res. *39*, D691–697.
Croucher, N.J., Harris S.R., Fraser, C., Quaill, M.A., Burton, J., van der Linden, M., McGee, L., von Gottberg, A., Song, J.H., Ko, K.S., *et al.* (2011) Rapid pneumococcal evolution in response to clinical interventions. Science *331*, 430–434.
Eldar, A., and Elowitz, M.B. (2010). Functional roles for noise in genetic circuits. Nature *467*, 167–173.
Galan, J.E. (2009). Common themes in the design and function of bacterial effectors. Cell Host Microbe *5*, 571–579.
Gawronski, J.D., Wong, S.M.S., Giannoukos, G., Ward, D.V., and Akerley, B.J. (2009). Tracking insertion mutants within libraries by deep sequencing and a genome-wide screen for *Haemophilus* genes required in the lung. Proc. Natl. Acad. Sci. U.S.A. *106*, 16422–16427.
Helaine, S., Thompson, J.A., Watson, K.G., Liu, M., Boyle, C., and Holden, D.W. (2010). Dynamics of intracellular bacterial replication at the single cell level. Proc. Natl. Acad. Sci. U.S.A. *107*, 3746–3751.
Keseler, I.M., Collado-Vides, J., Santos-Zavaleta, A., Peralta-Gil, M., Gama-Castro, S., Muniz-Rascado, L., Bonavides-Martinez, C., Paley, S., Krummenacker, M., Altman, T., *et al.* (2011). EcoCyc: a comprehensive database of *Escherichia coli* biology. Nucleic Acids Res. *39*, D583–D590.
Khan, S.M., Franke-Fayard, B., Mair, G.R., Lasonder, E., Janse, C.J., Mann, M, and Waters, A.P. (2005). Proteome analysis of separated male and female gametocytes reveals novel sex-specific *Plasmodium* biology. Cell *121*, 675–687.
Langridge, G.C., Phan, M.D., Turner, D.J., Perkins, T.T., Parts, L., Haase, J., Charles, I., Maskell, D.J., Peters, S.E., Dougan, G., *et al.* (2009). Simultaneous assay of every *Salmonella* Typhi gene using one million transposon mutants. Genome Res. *19*, 2308–2316.
Macek,B., Gnad, F., Soufi, B., Kumar, C., Olsen, J.V., Mijakovic, I., and Mann, M. (2008). Phosphoproteome analysis of *E. coli* reveals evolutionary conservation of bacterial Ser/Thr/Tyr phosphorylation. Mol. Cell. Proteomics 7, 299–307.
Malmstrom, J., Beck, M., Schmidt, A., Lange, V., Deutsch, E.W., and Aebersold, R. (2009). Proteome-wide cellular protein concentrations of the human pathogen *Leptospira interrogans*. Nature *460*, 762–765.
Mazurkiewicz, P., Tang, C.M., Boon, C., and Holden, D.W. (2006). Signature-tagged mutagenesis: barcoding mutants for genome-wide screens. Nat. Rev. Genet. *7*, 929–939.
Oberhardt, M.A., Palsson, B. O., and Papin, J.A. (2009). Applications of genome-scale metabolic reconstructions. Mol. Syst. Biol. *5*, 320.
Payne, D.J., Gwynn, M.N., Holmes, D.J., and Pompliano, D.L. (2007). Drugs for bad bugs: confronting the challenges of antibacterial discovery. Nat. Rev. Drug Discov. *6*, 29–40.
Sharma, C.M., Hoffmann, S., Darfeuille, F., Reignier, J., Findei, S., Sittka, A., Chabas, S., Reiche, K., Hackermüller, J., Reinhardt, R., *et al.* (2010). The primary transcriptome of the major human pathogen *Helicobacter pylori*. Nature *464*, 250–255.
Szklarczyk, D., Franceschini, A., Kuhn, M., Simonovic, M., Roth, A., Minguez, P., Doerks, T., Stark, M., Muller, J., Bork, P., *et al.* (2011). The STRING database in 2011: functional interaction networks of proteins, globally integrated and scored. Nucleic Acids Res. *39*, D561–D568.
Wang, D., and Bodovitz, S. (2010) Single cell analysis: the new frontier in 'omics'. Trends Biotechnol. *28*, 281–290.
Wang, K., Li, M., and Hakonarson, H. (2010). Analysing biological pathways in genome-wide association studies. Nat. Rev. Genet. *11*, 843–854.
Wang, Q., Zhang, Y., Yang, C., Xiong, H., Lin, Y., Yao, J., Li, H., Xie, L., Zhao, W., Yao, Y., *et al.* (2010). Acetylation of metabolic enzymes coordinates carbon source utilization and metabolic flux. Science *327*, 1004–1007.
Yoon, H., Ansong, C., McDermott, J.E., Gritsenko, M., Smith, R.D., Heffron, F., and Adkins, J.N. (2011). Systems analysis of multiple regulator perturbations allows discovery of virulence factors in *Salmonella*. BMC Syst. Biol. *5*, 100.

Manipulating the Fight Between Human Host Cells and Intracellular Pathogens

5

Rico Barsacchi, Varadharajan Sundaramurthy, Kees Korbee, Jacques Neefjes, Tom H.M. Ottenhoff, Tiziana Scanu and Marino Zerial

Abstract

Host-microbe interactions are complex phenomena spanning multiple levels of complexity, from environmental and ecological factors up to the cellular and genetic levels of host responses. At each of these levels a relationship is established between one or more micro-organisms and the host, resulting in formation of various forms of associations ranging from symbiosis to parasitism.

Pathogens have the potential to cause disease in their hosts through host–pathogen interactions in which host defences are challenged by the invasive capacities of the pathogen. The aim of this chapter is to give an outline of attempts made to unravel the components of host–pathogen interactions at the cellular and molecular levels and to discuss strategies to skew the balance in favour of the host, focusing our attention on crucial intracellular pathogens causing globally relevant diseases such as tuberculosis, gastroenteritis, influenza and malaria.

Host–pathogen relationships

Bacteria and viruses represent the major source of pathogens for human hosts, but eukaryotic organisms like helminths or protists can also cause diseases. Infectious agents get in contact with their particular host from the environment (or an in invertebrate vector) and once inside the body infect specific cell types or organs. These pathogens are the cause of major human diseases, leading to morbidity and mortality on a global scale.

Mycobacterium tuberculosis

Mycobacterium tuberculosis represents a critical health issue, with one-third of the human population carrying a latent infection and around 1.5 million annual deaths in HIV-negative cases alone (Diedrich and Flynn, 2011). *M. tuberculosis* is able to maintain infection through the entire life of the host in a dormant state, ready to be reactivated by as of yet largely unknown cues in normally immunocompentent individuals, or by immune suppression in immunocompromised patients (e.g. HIV patients).

A large part of the 'success' of *M. tuberculosis* in persisting in the host in a dormant state can be explained by its ability to exploit beneficial host cellular processes, blocking microbicidal ones and ultimately hijacking the host cell machinery to establish and maintain an intracellular survival niche.

After aerosol-mediated infection of the bronchoalveolar space *M. tuberculosis* is phagocytosed by alveolar macrophages (Kaufmann, 2001). The entry of *M. tuberculosis* into the macrophage takes place through redundant receptor-mediated systems in which a large array of different molecules, such as complement receptors, mannose receptors, C-type 2 lectin receptors (DC-SIGN, Mincle, Dectin-1), Fc receptors and integrins have been implicated (Brennan and Nikaido, 1995). The peculiar and very diverse constituents of the *M. tuberculosis* cell wall might explain the redundancy of this system. Upon entry into the host cell *M. tuberculosis* employs several different strategies to escape the normal macrophage innate immune functions. *M. tuberculosis* is able to interfere with the endocytic machinery by interfering with lipid signalling, by producing 'host-like' factors (such as kinases with mimicking effects), it can interact with the death-inducing pathways which are important for the establishment of immune reactions and bacterial clearance, it can inhibit normal calcium signalling functions, and it can inhibit macrophage activation. These are elaborated below. As a collective result of these events *M. tuberculosis* establishes a secure niche in which the bacterium is able to reside for decades.

The interaction of *M. tuberculosis* with host lipid signalling occurs at the level of phosphatidylinositol 3-phosphate (PI(3)P). This signalling intermediate is a central relay in the process of docking of early endosomes on the phagosome, which in turn is critical for the progression of phagosome maturation. The bacterium is able to interfere with the activity of the phosophatidylinositol 3-kinase (PI(3)K) VPS34, preventing the generation of PI(3)P on the phagosome membrane. This is possibly achieved through the action of the bacterial cell wall lipid lipoarabinomannan (LAM) (Vergne *et al.*, 2003), as well as by secretion of the acidic host-like phosphatase SapM, which is thought to hydrolyse PI(3)P on the phagosomal membrane (Saleh and Belisle, 2000). The combined effect of these events is an overall reduction of PI(3)P on the phagosome membrane, which results in the inhibition of phagosome maturation. In addition, the *M. tuberculosis*-containing phagosome can retain its ability to interact with early and recycling endosomes through the action of RabGTPases 11 and 14 (Kyei *et al.*, 2006). This is thought to provide the bacterium with access to nutrients circulating through the recycling endosomes.

Thus, *M. tuberculosis* blocks phagosome maturation progression both to avoid destructive late endosomal or lysosomal enzymes, and to remain able to access nutrients that the host is internalizing.

In addition to SapM, *M. tuberculosis* is able to express protein kinase G (PknG), a soluble eukaryotic-like serine-threonine kinase whose function is proposed to be phosphorylation of an as of yet unidentified host factor. The activity of this kinase is essential for inhibition of phagosome maturation by *M. tuberculosis* as intracellular trafficking of a PknG deletion mutant results in lysosomal delivery (Walburger *et al.*, 2004).

Another mechanism that *M. tuberculosis* employs to modulate host defences is perturbation of calcium signalling. Recruitment of the host protein coronin-1/TACO to the mycobacterial phagosome is critical for arresting phagosome maturation. Mycobacteria induce cytosolic calcium flux in a coronin-1-dependent manner that results in activation of the calcium-dependent phosphatase calcineurin. Thus, genetic ablation of coronin-1 (Jayachandran *et al.*, 2007, 2008) or chemical inhibition of calcineurin (Jayachandran *et al.*, 2007) results in release of mycobacterium-mediated phagosome maturation arrest *in vitro* and *in vivo* and subsequent killing of intracellular mycobacteria *in vitro*.

Recent studies demonstrated that *M. tuberculosis* actively manipulates the innate immune response of the host in order to promote its survival and immune escape through several different mechanisms. First of all, *M. tuberculosis* induces decreased secretion of the pro-inflammatory cytokine IL-12. Nau *et al.* demonstrated this in 2002 by analysis of gene expression profiles of human macrophages upon exposure to a panel of different bacterial strains and bacterial components. By analysing *M. tuberculosis*-infection specific differentially regulated genes a marked decrease in expression of IL-12 was observed in response to infection with this pathogen. The observation that IL-12 was down-regulated specifically by *M. tuberculosis* but not by other bacteria was an indication that *M. tuberculosis* might modulate other host signalling pathways in addition to those involved in Toll-like receptor (TLR) signalling (Nau *et al.*, 2002). Secondly, *M. tuberculosis* interferes with signalling pathways downstream of the IFNγ receptor (IFNγR), thus inhibiting a major protective cytokine pathway in human host defence to mycobacteria (Ottenhoff *et al.*, 2002; Fortune *et al.*, 2004; Banaiee *et al.*, 2006; Benson and Ernst, 2009; Desvignes and Ernst, 2009). A third mechanism is down-regulation of HLA class II molecules at the macrophage cell surface to subvert $CD4^+$ T-cell responses (Ting *et al.*, 1999; Noss *et al.*, 2001; Kincaid and Ernst, 2003; Diedrich and Flynn, 2011). Additionally, *M. tuberculosis* targets DC-SIGN (CD209), which has been shown to deactivate infected DCs and induce IL-10 production in specific settings (Appelmelk *et al.*, 2003); this may promote immune deviation from Th1 type immunity. *M. tuberculosis* also actively induces regulatory T cells, and the secretion of TGFβ and IL-10, all of which are involved in dampening host immunity.

In recent years, new severe health risks related to *M. tuberculosis* have emerged that have dramatically increased the urgency to find new drugs and targets against tuberculosis. First, HIV co-infection enhances reactivation of latent *M. tuberculosis* and in turn *M. tuberculosis* could possibly augment the severity of disease in HIV-infected individuals (Diedrich and Flynn, 2011). HIV infects about one-fifth of people with *M. tuberculosis* infection in southern Africa and TB is now the leading cause of mortality in HIV-infected patients (Kaufmann and Parida, 2008). The magnitude of severity is emphasized by the fact that a HIV infected person has 20-fold more chance to develop active TB (Jassal and Bishai, 2010). Second, multidrug-resistant (MDR) and extensively drug-resistant (XDR) *M. tuberculosis* strains have been emerging (Jassal and Bishai, 2010), underlining the necessity for new therapeutic approaches, either through novel antibiotics or by host cell approaches. However, the scarcity of available targets and lack of an integrative understanding of the host pathways manipulated by *M. tuberculosis* imposes a considerable bottleneck for research on novel therapeutic approaches (Meresse *et al.*, 1999; Vergne *et al.*, 2004). Therefore, the field of tuberculosis research has seen the emergence of large-scale screening and systems biology approaches to aid in the identification of new targets in both the host and the pathogen.

Salmonella

Salmonella enterica is another example of an intracellular bacterial pathogen that is successful in exploiting host functions to find an intracellular niche in which it is able to proliferate and perpetuate the infection. *Salmonella enterica* serovars Typhi (*S. typhi*) and Paratyphi (*S. paratyphi*) are facultative intracellular bacteria that cause gastroenteritis and typhoid fever, a chronic, systemic illness with symptoms ranging from fever, abdominal pain and rash to possible transient diabetes, and a mortality rate of 10–15% if left untreated. In contrast, *Salmonella enterica* serovar Typhimurium (*S. typhimurium*) is unable to cause typhoid fever in humans, but is a common causative agent of gastroenteritis. However, infection of a murine host with this pathogen can result in a systemic disease resembling human typhoid fever.

Other *Salmonella* species, like *S. enteritidis*, are able to infect a wide range of other animals and cause transient enteritis in humans.

Salmonella is generally contracted through the ingestion of contaminated food or water. The first line of contact with the host is the stomach, where the low gastric pH represents the initial defence against infection. After entering the small bowel, *Salmonella* must traverse the intestinal mucus, where it induces fluid and ion secretion into the lumen of the bowel as well as an acute inflammatory reaction (Santos *et al.*, 2001).

The first level of cellular interaction with the host is represented by the contact with the intestinal epithelium. The overall strategy that the pathogen has evolved is to induce phagocyte-like behaviour in otherwise non-phagocytic competent cell types, like the intestinal epithelium. This process involves exploitation of host signalling pathways, trafficking systems and most importantly regulation of the actin cytoskeleton (Guiney and Lesnick, 2005).

The ability to invade and populate the epithelial cell layer depends on a genetic locus termed the *Salmonella* pathogenicity island 1 (SPI1), which encodes a type III protein secretion system that is able to deliver a series of bacterial effector proteins to the cytoplasm of the host cell. Effectors coded by SPI1 known to be responsible for manipulation of the host cell machinery are SopB (an inositol 3-phosphatase), SopE and SopE2 (both guanine nucleotide exchange factors for Cdc42 and RAC1), which are activators of the Rho family GTPases Cdc42 and Rac1 (Hardt *et al.*, 1998). This activation leads to the recruitment of WASP and Scar/WAVE family proteins, which together with Arp2/3 complexes initiate actin polymerization and permit the formation of the *Salmonella*-containing vesicle (SCV) in the non-phagocytic host cell population.

Once formed, the SCV undergoes a maturation process by interacting transiently first with early endosomes and then with late endosomes (Brumell and Grinstein, 2004). During this maturation process, the fusion of the SCV with lysosomes is thought to be prevented, allowing *Salmonella* to survive. The SCV can however acquire and retain some lysosomal membrane markers like the glycoprotein LAMP-1 (Holden, 2002).

Salmonella modulates the host's microtubular kinesin motors on phagosomes and manipulates dynein motor-driven transport of phagosomes to lysosomes to rescue *Salmonella* growth (Harrison *et al.*, 2003; Marsman *et al.*, 2004; Boucrot *et al.*, 2005). Moreover, *Salmonella* has been shown to activate host kinases, which might inhibit dynein motors to promote *Salmonella* replication (Uchiya *et al.*, 1999).

After invasion, the intracellular phase is followed by passage into the subepithelial interstitium, where phagocytosis-competent neutrophils and macrophages are able to ingest the pathogen. Here, the bacterium is able to survive in non-activated macrophages, resulting in systemic spread and dissemination to various organs like regional lymph nodes, the liver and the spleen.

The intracellular macrophage phase of *Salmonella* is characterized again by the processing of the SCV in a unique route of intracellular vesicular trafficking. In this case, the exploitation of host mechanisms is facilitated by the expression of a second genetic locus in the pathogen genome, called *Salmonella* pathogenicity island 2 (SPI2). SPI2 codes for another type III protein secretion system, which function is to transfuse virulence factors from the pathogen to the host through the vacuolar membrane. The SPI2 locus codes for effectors that diffuse into the host cell cytoplasm, altering its physiology through formation of filamentous endosomes (Stein *et al.*, 1996), decreased recruitment of NADPH oxidase (Vazquez-Torres *et al.*, 2000) and changes in the SCV fusion pathways (Uchiya *et al.*, 1999, 1999), as well as by controlling the actin cytoskeleton in close proximity to the SCV.

One of the survival strategies of *Salmonella* is exerted through the action of the bacterial SpvB protein, which counteracts actin polymerization and is involved in triggering apoptosis. *In vitro*, macrophages infected with *Salmonella* display a profound loss of actin filaments at 10 hours post infection. After 20–24 hours many of these cells have no more detectable polymerized actin and undergo DNA fragmentation in a process resembling apoptosis. This is thought to permit the SCV to exit the cell with little exposure to the extracellular environment, allowing subsequent cycles of infections. Necrosis is probably not involved in dissemination of the bacteria as the vacuoles released by that process would lose their structural integrity and the bacterium would be exposed to the unfavourable extracellular milieu.

An example of viral infection: influenza

In addition to prokaryotic infections, new or re-emerging viral pathogens also represent a global disease issue. Influenza virus represents a prototypic class, being the most common cause of human respiratory infection and mortality.

Influenza virus is an enveloped negative-strand segmented RNA virus, which represents three of the five genera of the *Orthomyxoviridae*. The classification of the virus depends mainly on the host tropism and pathogenicity: influenza virus A, which natural hosts are wild aquatic birds, influenza virus B, which almost exclusively infects humans, and the rarest, influenza virus C, which infects mostly humans, dogs and pigs (Taubenberger and Kash, 2010).

Cross-species transmission is possible and, in the case of influenza virus A, has been responsible for outbreaks in the poultry population or in human pandemics. Examples of these were the avian flu outbreak in 2004 or the swine flu outbreak in 2009.

The viral genome, consisting of eight RNA segments coding for ten viral proteins, is enveloped in a host-derived lipid membrane (Lenard, 1978). Three viral proteins are embedded in the lipid membrane: haemagglutinin (HA), whose major function is binding to and recognition by host receptors, neuraminidase (NA), a glycoprotein with sialidase activity important for the release of newly produced virus, and a minor number of matrix 2 (M2) ion channel proteins required for viral replication. The viral core contains eight viral ribonucleoprotein (vRNP) complexes, each one composed of the viral RNA together with the nucleoprotein NP and components of the viral polymerase complex, namely PB2, PB1 and PA.

Infection is initiated by viral attachment to cells via binding of HA proteins to sialic acid (SA) surface receptors on the host cell. The multiplicity of HA receptors on the viral capsid surface reinforces the otherwise weak binding (Matrosovich *et al.*, 2009). The HA protein is critical for host specificity, displaying highly conserved regions in the receptor binding domain of the HA protein of human-specific viruses. In the acidic environment of the endosomes, vRNPs are released into the cytoplasm where they will have to cross the nuclear barrier before entering the nucleus, their final destination. The transport across the nuclear barrier is mediated by host importins. Mutations of the PB2 polymerase subunit and the nucleoprotein (NP) have been shown to increase the binding to the host importin-alpha, hence promoting virus transcription and replication in the nucleus (Gabriel *et al.*, 2008).

Importin is known to act as an adaptor protein that recognizes the nuclear localization signal (NLS) of the cargo protein. This complex is then transported with the importin beta receptor into the nucleus, effectively hijacking the host system to enter the nucleus and replicate.

Once inside the nucleus the influenza polymerases are able to replicate the vRNA. The influenza polymerase is a low quality polymerase without proofreading activity, hence prone to introduction of errors in replication. This is a crucial determinant in the evolutionary strategy of the virus, as a continuous change in the glycoprotein sequence is a key aspect of evading host antibody response (the so called 'antigenic drift').

Following the production of new vRNA, the viral genome is encapsulated into new capsid proteins and trafficked to areas of the plasma membrane where envelope HA and NA have been inserted. Following assembly, the mature virion is released by membrane budding, ready to perpetuate the infection cycle.

From this description, it is immediately clear that a small group of viral proteins requires interaction with several cellular components at various levels, with several cellular functions. While the knowledge of viral proteins is advanced to a certain degree, relatively less is known about the network of interactions with cellular host factors.

Several systematic screens have been performed in the last years in this context, generating a wealth of information. To rationalize and integrate the knowledge from each of these attempts, some examples of systematic screens will be discussed in the second section of the chapter.

Malaria strategies

Plasmodium, the parasite responsible for malaria in humans, is an example of a eukaryotic parasite that has developed a complex life cycle involving the obligate passage through more than one host. Following an infected anopheles mosquito bite, the *Plasmodium* sporozoites are injected in the bloodstream and travel through it to reach the liver, where they colonize hepatocytes, divide asexually and mature into schizoints containing many merozoites. The liver stage is the clinically silent bottleneck phase of the disease that is necessary for the parasite to cross in order to successfully replicate and disseminate. In this stage the small amount of sporozoites injected by an infected mosquito rapidly develop into hundreds of thousands of merozoites that are able to invade erythrocytes. It is in these cells that a synchronized division phase takes place, ultimately causing clinical symptoms of malaria such as periodic high fever and chills. A proportion of the parasites can mature at this stage into gametocytes, which are transferred back into a mosquito upon a bite. The sexual combination occurs in the bowel of the mosquito, forming a diploid zygote. The zygotes differentiate in the epithelial wall of the bowel, developing into oocysts, which subsequently undergo mitotic division to produce new sporozoites. These then migrate to the salivary gland of the mosquito, awaiting injection into a new human host, perpetuating the cycle. In the context of this complex life cycle a large series of interactions takes place between the parasite and the host.

Once inside the erythrocytes, it is essential for the *Plasmodium* parasites to acquire nutrients. One strategy utilized by the parasite is the degradation of haemoglobin. Interestingly, several polymorphisms in the haemoglobin genes are present in the human population that, although causing mild to severe anaemia, can endow individuals with protection against *Plasmodium* infection (Weatherall, 2008; Bongfen *et al.*, 2009). A particular and famous case of protection to *Plasmodium* infection can be found in thalassaemias, caused by heterozygosity for null alleles of isoforms of haemoglobin A and B.

The digestion of haemoglobin by *Plasmodium* causes oxidative stress and changes in the redox potential of the erythrocyte. Glucose-6-phosphate dehydrogenase (G6PD) protects against this stress by generating $NADPH^+$, which is required to produce glutathione. There is genetic evidence that G6PD deficiency can be linked to protection against *Plasmodium* infection and invasion, due to the high disruption rate of erythrocytes upon infection (Cappellini and Fiorelli, 2008).

These examples in general reflect an interesting selective pressure imposed by the pathogen in order to partially inactivate important housekeeping genes of the host. Partial failure of the proteins encoded by these genes could result in beneficial effects on the pathogenicity of the *Plasmodium* parasite. In these cases the host is genetically under selective pressure where certain pathologies can give rise to protection against the pathogen in a genetic 'tug-of-war'.

The completion of the life cycle of *Plasmodium*, the sexual phase, is spatially confined to the midgut of the Anopheles mosquito after ingestion of infected blood. The insect and *Plasmodium* factors involved in the recognition and/or elimination of the pathogen remain largely unknown. This is in part due to the difficulty of developing experimental models for this stage of the life cycle of the parasite and the scarcity of available genetic data (Bongfen *et al.*, 2009).

Malaria is also is an interesting showcase for host genetic adaptation to disease. Host genetic variants that highlight vulnerable spots in the ability of the parasite to replicate and invade the host have been identified, like the Duffy negativity or the Gerbich erythrocyte variants. In the first case there is an absence of expression of the Duffy antigen receptor for chemokines (Langhi and Bordin, 2006) in the erythrocytes, which causes a resistance to invasion by *Plasmodium vivax* (Langhi and Bordin, 2006). The second case refers to the lack of expression of a series of glycophorins in erythrocytes leading to impaired *Plasmodium* invasion (Maier *et al.*, 2003).

Targeting the host to fight pathogens

As described in the first part of this chapter, intracellular microbes interact extensively with their host environments and modulate it to suit their own survival, often resulting in a shift of the equilibrium of several host functional modules. Hence, these interactions can be considered as perturbations favoured by evolution to facilitate pathogen survival.

Comprehensive understanding of the nature of these dynamic interactions might give us new targets to confront the pathogen and enable us to re-shift the equilibrium in favour of the host. In order to do so, novel tools to dissect the nature of these perturbations and to integrate them into functional networks are essential. Several systematic approaches in the form of high throughput functional screens have been used in recent years to explore the interface between host and pathogen in the previously described cases. A computational analysis of interaction nodes of known and experimentally validated pathogen proteins and their host interactors demonstrated that the host molecules that are most often targeted also tend to form crucial nodes in various host pathways. Pathogens tend to target host molecules in a highly directed manner, specifically influencing components that are essential for the host cell and hence are not easily modifiable by the host in order to overcome the influence of pathogen (Dyer *et al.*, 2008). However, this could be turned to the advantage of the host in a therapeutic setting, since these nodes are usually well characterized. Hence, a wealth of information exists on the biology and pharmacology of these targets, which can potentially be employed in the fight against pathogens. Moreover, targeting the host to fight pathogens might have the additional advantage of reducing chances of drug resistance. The chances of a pathogen acquiring resistance to a host process are arguably lower than classical microbial drug resistance, since in the former case no direct selective pressure is exerted on the pathogen (Schwegmann and Brombacher, 2008). Hence an emerging approach in the field is to target host processes to fight pathogens.

Influenza

There has been significant activity in recent years performing screens for host factors important in Influenza virus infection, with at least five independent screens published from 2008 until now (Hao *et al.*, 2008; Brass *et al.*, 2009; Shapira *et al.*, 2009; Sui *et al.*, 2009; Karlas *et al.*, 2010; Konig *et al.*, 2010). Owing to differences in the methodologies, cell systems, screening methods, analysis tools and the complexity of the system as a whole, there was little overlap in the hits obtained in these screens. Although it is possible that a broad space encompassing all hits from all screens could cover a spectrum of host determinants of influenza infection, more plausible reasoning is that a restricted spectrum of overlapping hits could define a minimal set of host determinants of influenza infection. There have been several attempts to integrate data from these screens to get a comprehensive overview of host factors involved in the influenza virus life cycle (Watanabe *et al.*, 2010; Stertz and Shaw, 2011). Sterz and Shaw have listed the primary hits of each of these screens separately from the validated hits and considered a gene validated if at least two siRNAs resulted in a significant reduction of viral replication. As a result a total of 478 genes were validated and 34 genes were shared between at least two screens. Within the components of this list two groups of genes were highly enriched: one encoding the vacuolar ATPase (vATPase) subunits and one encoding the COPI proteins. Vacuolar ATPase is required for endosome acidification and influenza virus is dependent on this process for invasion of host cells. The COPI proteins are known to mediate retrograde transport from the Golgi to the endoplasmic reticulum under physiological conditions. However, the function of these proteins in the context of viral infection requires further investigation. Other nodes identified comprised genes involved in pre-mRNA splicing, which are possibly hijacked to allow viral mRNA splicing, and genes involved in nuclear transport of mRNA and proteins (Stertz and Shaw, 2011).

Using a similar approach, Watanabe and colleagues (2010) analysed the results of each screen and compared the validated lists of hits pairwise, identifying a set 128 genes overlapping in at least two of the screens. This knowledge was integrated with data on host protein–virus interaction and viral protein interactions with the aim of mapping the viral proteins on the genes of interest and vice versa. When this network was probed for biological processes, the enriched nodes were highly similar to those mentioned above: endocytosis (e.g. components of the coatomer complex or subunits of vATPase), nuclear transport, pre-mRNA processing and translation initiation.

The results of these analyses, although probably still neither conclusive nor exhaustive, represents a starting point from where the interaction of viral proteins with host factors can be more systematically evaluated and possibly exploited from a therapeutic point of view.

Tuberculosis

Although many interactions between viral and host proteins are documented and validated experimentally, the same cannot be said for bacteria and parasites. In fact, even for relatively well-studied bacteria like *M. tuberculosis* there are only few direct interactions mapped between bacterial proteins and their host interactors (Bach *et al.*, 2008; Volkman *et al.*, 2010). Previous studies from different labs have identified several host molecules involved in mycobacterial infection, which were verified by either RNAi-mediated gene silencing and/or chemical compound treatment. Inhibition of the identified proteins restored intracellular balance in favour of the host. Examples of perturbations of the host system known to induce intracellular killing of mycobacteria are RNAi against host coronin (Jayachandran *et al.*, 2008), chemical inhibition of calcineurin (an effector of coronin 1) (Jayachandran *et al.*, 2007), RNAi against host Rab14 (Kyei *et al.*, 2006) and treatment of cells with AX20017 (an inhibitor of bacterial PknG that is secreted in the host cell upon infection and is thought to modulate host proteins) (Walburger *et al.*, 2004). Despite these isolated examples, system-wide understanding of interactions between mycobacteria and host cells is incomplete and therefore inadequate to rationally or systematically perturb host systems for therapeutic purposes. In an effort to fill this crucial gap, several laboratories are currently employing new technologies like arrayed RNAi or chemical compound libraries to probe the system and are using techniques such as automated microscopy, high content analysis, proteomics, microarrays and deep sequencing to visualize and analyse interactions between host and pathogen.

In the work of C. Kuijl and co-workers (Kuijl *et al.*, 2007), several host kinases capable of inhibiting intracellular growth of *S. typhimurium* were identified through an RNA interference screen targeting the human kinome using automated fluorescence microscopy. In particular, the kinase PKB/Akt1 was identified as an important regulator of phagosome maturation. Akt1 is activated by *Salmonella* protein SopB, which is actively secreted into the host cytosol through the bacterial type III secretion system (TTSS).

S. typhimurium activates Akt1 for several reasons: (1) Akt1 activation provides a survival signal for the cells, affording a continued niche for bacterial survival; (2) Akt1 ultimately activates the host kinase PAK4 which controls Rac1 activation and actin polymerization (Kuijl *et al.*, 2007). Actin plays an important role in the maintenance of phagosomes; (3) Akt1 heavily phosphorylates the RAB14-GAP, AS160, which subsequently dissociates from the phagosomal membrane, thereby preventing AS160s ability to inactivate the GTPase Rab14. Active Rab14 prevents fusion of phagosomes with lysosomes and thereby promotes *S. typhimurium* survival. Of note, the same pathway appears to be activated by a series of unrelated intracellular bacteria such as *M. tuberculosis*, *Shigella* and *Chlamydia* strains (Kuijl *et al.*, 2007).

Thus, activation of Akt1 appears to be a general mechanism exploited by intracellular pathogens to manipulate the host to facilitate pathogen survival. A chemical inhibitor of Akt1 called H-89 used in the study by Kuijl *et al.* was shown to be effective in decreasing intracellular survival of both *S. typhimurium* and *M. tuberculosis* in human primary macrophages (Mφs). This indicated that similar signalling pathways might also play an important regulatory role in intracellular trafficking of *M. tuberculosis*. Inhibition of Akt1 may therefore be a new strategy to eliminate intracellular bacteria. In addition, multidrug resistant and extensively drug-resistant strains of mycobacteria such as *M. tuberculosis* might still be susceptible to the effects of inhibition of Akt1.

Rao and colleagues performed two screens aimed to identify host effectors of intracellular mycobacterial infection and identified several host molecules that are crucial for bacterial infection. Kumar *et al.* (2010) reported a genome-wide screen to identify host factors involved in the regulation of the bacterial load of a virulent *M. tuberculosis* strain (H37Rv) in the THP-1 human monocytic leukaemia cell line (Kumar *et al.*, 2010). In order to specifically study host factors involved in intracellular replication or degradation but not uptake of *M. tuberculosis*, siRNA transfections were performed post infection, in contrast to the approach by Kuijl *et al.* (2007). Using this post-infection siRNA approach the authors were unable to confirm a major role for Akt1, but silencing of Akt1 *prior* to infection resulted in a ~50% decrease in *M. tuberculosis* bacterial load, in agreement with the findings of Kuijl *et al.* The genome-wide screen of Kumar *et al.* (2007) yielded an array of 275 host factors that influenced *M. tuberculosis* bacterial load in THP-1 cells, from which an '*M. tuberculosis* regulatory network' was constructed.

This network of interactions was further subclustered based on gene ontology annotations (Meresse *et al.*, 1999). Interestingly, several subnetworks in which the GO classes 'immunity/inflammation/stress', 'transcription regulation', 'signal transduction', 'metabolism' and 'cell cycle/growth' were enriched centred around Akt1, further emphasizing the importance of this kinase in the regulation of *M. tuberculosis* intracellular survival and replication. By silencing the 275 host factors identified in cells infected with several different *M. tuberculosis* field isolates (each with different virulence, antibiotic resistance and intracellular growth properties), 74 'common' host factors that were *M. tuberculosis* strain-independent were identified.

Based on the surprising observation that diverse field isolates of *M. tuberculosis* respond differently to host perturbations, one can argue that targeted host-based therapies must account for the heterogeneity of host signatures for different mycobacterial infections. Despite this plasticity, there appears to be a core component of host responses that could potentially constitute a minimal module of host defence, as defined by the identification of strain-independent factors. Within this subset the effect of 44 siRNA pools could be reversed by chemical inhibition of autophagy, indicating that these host factors might be involved in the negative regulation of autophagy. Although the anti-mycobacterial effects of autophagy have previously been shown (Gutierrez *et al.*, 2004), the extent of its interaction with mycobacterial infections was not fully appreciated. This study suggests that inhibition of autophagy might be a central axis of mycobacterial intracellular survival.

Since it emerges that mycobacteria actively inhibit autophagy, novel inducers of autophagy identified via independent chemical screens can be tested for their potential use as anti-mycobacterial agents, thereby opening new avenues in treatment of tuberculosis. Identification of other critical host survival axes could therefore be important in devising targeted host based therapies. Another interesting observation was that IFN-γ stimulation appeared to enhance the suppressive effect of the silencing of several host factors on *M. tuberculosis* intracellular replication, despite the fact that the infected cells were relatively insensitive to IFN-γ (Beaulieu *et al.*, 2010; Kumar *et al.*, 2010).

Jayaswal *et al.* published another study in 2010 employing similar methods to those used by Kumar *et al.* (2010). The authors reported an RNAi screen focused on the kinome and phosphatome in the J774 murine macrophage cell line, resulting in the identification of 41 host factors involved in the intracellular replication of *M. tuberculosis*. Using a similar approach as in Kumar *et al.* (2010), a selection of 11 host factors was found to be *M. tuberculosis* strain independent. From this selection, the TGFβ type 1 receptor (TGFβRI) was studied in more detail, since the expression of both this receptor and another isoform (TGFβ type 2 receptor, TGFβRII) as well as the ligands TGFβI and TGFβII was found to be up-regulated in response to *M. tuberculosis* infection of mouse primary macrophages by microarray analysis. In addition, casein kinase 1 (CSNK1) was studied, which is a kinase in the TGFβR1 axis and one of the 11 identified *M. tuberculosis* strain-independent host factors. A chemical inhibitor of both TGFβRI and CSNK1 was found to be effective in decreasing bacterial load after infection of mouse primary macrophages with H37Rv or two different clinical field isolates. In addition, it was also effective in a murine *M. tuberculosis* infection model (Jayaswal *et al.*, 2010), as assessed by colony-forming unit (CFU) counts and histopathology of infected organs. This provides direct evidence of the feasibility of host-based therapeutic approaches to target mycobacteria.

A different approach to better understand the intersection between *M. tuberculosis* and its host has been presented recently by Nathan and colleagues (Beaulieu *et al.*, 2010). They have built a matrix of host effectors and bacterial mutants and screened this combinatorial space for phenotypes using an image based screening platform. Their system was geared to identify mycobacterial mutants that regulate host immunity and hence they engineered host cells to report on 12 different immune effectors of mycobacterial infections and scanned the whole mycobacterial genome for responses to these effectors. Despite the significant technical challenges in building up this matrix and the undefined rates of false positives, they have identified several new *M. tuberculosis* genes that play a role in manipulating host immunity and more than 200 genes as candidates for regulating the responses.

Such approaches using various defined host and pathogen gene classes to detail the intersection between host and pathogen in a high-throughput or high-content manner could pave the way for a better understanding of the dynamics of the interactions at the host–pathogen interface at a systems level. These studies have therefore provided a much needed first step towards the systematic identification of genes involved in mycobacterium–host interactions. However, given the varied responses usually obtained between screens conducted in different experimental conditions using various methodologies, and the diverse criteria used to define 'hits', it will be important to understand the degree of correlation of data obtained by these and future screening campaigns. Much work remains to be done to come to an integrative understanding of the pathways regulating intracellular survival of mycobacteria and the interplay between host and pathogen that defines the generation, maintenance and eventual outcome of *M. tuberculosis* infection.

Malaria liver stages

Screening-based approaches to understand host factors have also been applied to complex host–pathogen interactions like the liver stage of malaria.

The liver stage is the clinically silent bottle-neck phase of the disease. In this stage sporozoites injected by an infected mosquito develop into hundreds of thousands of merozoites over several days, achieving one of the highest growth rates for eukaryotic cells. Such rapid growth necessitates massive assimilation of building blocks such as lipids for membrane formation, and amino acids for ribosome biogenesis to fuel protein synthesis. Therefore it is essential for the parasite to keep the host in a hospitable state during this period of rapid growth. The parasite interacts extensively with the host during this stage, but knowledge of the nature of these interactions is limited, owing largely to the experimental refractoriness of the system.

In two independent screens, Mota and colleagues (Prudencio *et al.*, 2008; Rodrigues *et al.*, 2008) silenced expression of different classes of host proteins, kinases and lipoproteins, in a hepatocyte cell line. These cells were infected with fluorescently labelled *Plasmodium berghei* and monitored by automated microscopy with subsequent automated image analysis.

Filtering the data for hits resulted in the identification of a single kinase (Prudencio *et al.*, 2008) and a host scavenger receptor SR-B1 (Rodrigues *et al.*, 2008). Using siRNA injection in mice the authors demonstrated that silencing the corresponding genes *in vivo* indeed reduced the infection burden in mice. Moreover, inhibition of SR-B1 by various chemicals reduced the parasite load *in vitro* in hepatocytes infected with *P. berghei* as well as the human pathogen *P. falciparum*, and *in vivo* in the murine model. In another study, the same group identified host haem oxygenase (HO) as an important molecule involved in control of malarial infection (Epiphanio *et al.*, 2008). Chemical inhibition of this enzyme led to clearance of the parasite and cure of mice challenged with malaria.

Although the application of siRNA as a therapeutic vehicle is not fully realized, the approach nevertheless shows that it is indeed feasible to counter complex diseases such as malaria by targeting a single host factor that plays crucial role in a critical phase of the disease.

Chemical genetics of host–pathogen interactions

Identifying host components involved in pathogenesis by RNAi-based screens as outlined above, is a way of finding host targets and processes involved in pathogenesis. A complementary approach is to identify chemical entities that perturb the host–pathogen system and from this knowledge arrive at targets involved in pathogenesis (Fig. 5.1). This is an emerging area and several efforts are currently under way to address this crucial gap.

Literature in this field is evolving and several recent examples of different methodologies adopted to identify intracellular targets of hit compounds will be outlined below. Although the methods described here have been used in the context of pathogen factors, many of them can be potentially adapted to identify host targets as well.

One of the direct approaches to identify the cellular target of a lead compound is exemplified by the efforts of Pethe *et al.* (2010) in the context of tuberculosis. In this study a class of pyrimidine-imidazole (PI) compounds was identified that killed *M. tuberculosis* in a whole cell screen but was ineffective in an *in vivo* infection model. To understand the underlying mechanisms, spontaneous mutants of *M. tuberculosis* that were resistant to three of the lead compounds were isolated. Sequencing of the resistant clones revealed a single polymorphism in the gene Rv3696c, which encodes a glycerol kinase. Moreover, *M. tuberculosis* mutants that are deficient for this gene display increased resistance to PI compounds and failed to grow *in vitro* when glycerol was used as the sole carbon source. However, since the drug is not efficacious *in vivo*, the authors argue that glycerol is not the primary source of carbon for *M. tuberculosis in vivo*.

In an alternative approach, the authors used an affinity chromatography approach to identify possible cellular targets of another compound by incubating *M. tuberculosis* lysate

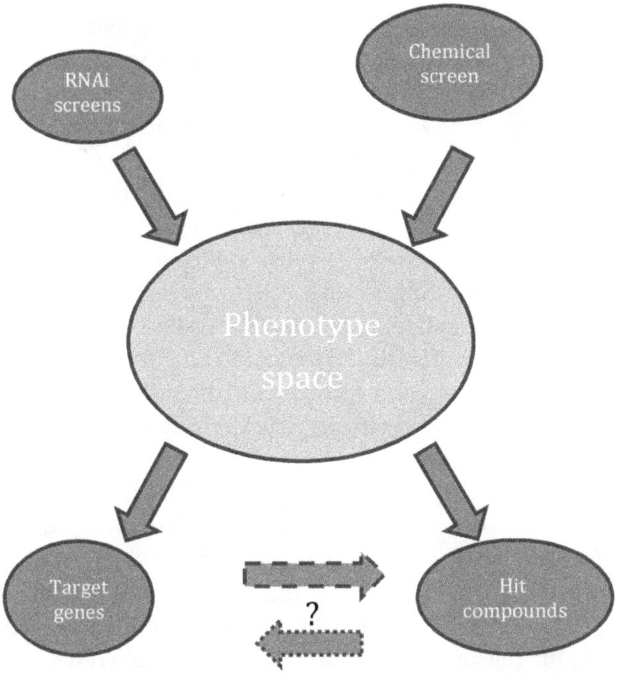

Figure 5.1 siRNA and chemical libraries have been used as perturbation agents to explore the phenotype space. The relationships between the genes and chemical entities that influence the phenotype space in a similar manner is relatively underexplored and might provide insights into biological mechanisms governing phenotypes as well as specific tools to study them.

with sepharose beads covalently linked to the compound of interest. Mass spectrometry of the resulting eluate identified four candidate proteins involved in a methylglyoxal detoxification pathway. Overexpression of Rv0557, a gene coding for one of the proteins, enabled *M. tuberculosis* to grow faster than wild-type bacteria on glycerol-containing media, suggesting that this gene might be involved in detoxification of glycerol metabolism by-products. Thus, by combining chemical screens, genetic approaches and biochemistry, the authors identified targets for the hit compounds, which then allowed them to understand a crucial aspect of the physiology of mycobacterial infection *in vivo*.

Recently, two papers were published that vastly increased the publicly available repertoire of known anti-malarial compounds (Gamo *et al.*, 2010; Guiguemde *et al.*, 2010). In addition, these papers reported interesting strategies to identify the mode of action of chemical compounds. Gamo *et al.* (2010) screened approximately 2 million chemical compounds for their anti-malarial effect. The hit compounds were clustered based on their chemical structure and by following a similarity principle the authors reasoned that compounds within the same structural cluster might share a similar mode of action. In an effort to explore these compounds vis-à-vis the known target space, the authors profiled representative compounds from different clusters based on their behaviour in diverse assays for different known 'classical' human drug targets. Such profiling allowed them to classify the hits based on their 'targets'. The authors then identified *P. falciparum* (Pfal) orthologues of these human targets and proposed a list of Pfal genes that might be considered as targets. Additionally, several compounds were identified that have no known orthologues in Pfal. Detailed studies on these might indeed help uncover previously under appreciated aspects of malarial physiology in general, and more specifically in the context of drug development.

In an alternative approach, Guiguemde *et al.* (2010) screened approximately 400,000 compounds for anti-malarial activity in a whole cell-based screen and identified and validated several of them as 'hits'. To understand the mode of action of the hit compounds, the authors assessed the performance of 66 selected hits in assays for known anti-malarial mechanism such as *P. falciparum* dihydro-orotate dehydrogenase (PFDHOD), *P. falciparum* falcipain-2 and hemazoin formation. Additionally, binding studies of these drugs were performed on a panel of 61 recombinant PF proteins. Using this approach, several combinations of possible mechanisms of action were identified for most of the drugs. Interestingly, five compounds were negative in all the assays tested despite having anti-malarial properties, thus showing novel mechanistic promise.

Volkman *et al.* (2010) proposed a different approach for identifying host cell targets for diverse perturbations such as toxins, drugs and viruses. In this approach gene trap retroviruses were used to knock out genes, generating a library of non-essential null mutants in the human near-haploid cell line 7KBM7. This library of random knock-out clones was then screened for the effects of the different knock-outs by applying different agents that are normally toxic to the cell line. While most clones will die as a result of this perturbation, surviving clones are likely to be rendered resistant by disruption of the genes coding for the targets of the toxic agents by insertion of retroviral DNA. This approach was successfully validated by treating mutated cells with Gleevec, an anti-cancer drug that specifically inhibits tyrosine kinase activity of ABL (and the fusion protein BCR–ABL that is constitutively active in certain tumour cells). As 7KBM7 cells are sensitive to Gleevec, surviving clones are likely to harbour an insertion in their target genes. Indeed, five independent Gleevec-resistant clones contained insertions in genes closely functionally related to c-ABL. In this approach genes are functionally knocked out (due to insertional mutagenesis) as opposed to knocked down in RNAi screens. Therefore it could complement RNAi-based screens by overcoming their significant drawbacks such as incomplete and heterologous silencing between different oligos and off-target effects. However, this tool is currently restricted to a particular cell line and therefore the biological processes supported by it, potentially limiting the exciting possibilities that it offers.

Conclusion

Functional and chemical genomics screens are powerful tools, but like every technology they are not free of caveats. Knockdown efficiency in a functional genomic screening is not homogeneous due to differences in efficacy between siRNA oligos, differences in mRNA and protein stability between targets, and possible redundancy of targets for which loss-of-function might be compensated by other factors in a single gene knockdown setting. On the other side chemical genomic screens are limited by other components, such as bioavailability, the use of a narrow window of concentrations (at least in the primary campaigns) or degradation of the compounds. Moreover, the analysis of the results poses both traditional and new challenges, with lack of standards for statistical and bioinformatic analysis and varying quality of different databases for accurate gene annotations, chemical–protein or protein–protein interactions. The availability of new approaches such as multiparametric analysis (Collinet *et al.*, 2010) and population context analysis (Snijder *et al.*, 2009) to evaluate phenotypes with a finer granularity, as well as development of tools for correlation analysis across different screens (as described for example in the influenza section), will certainly help to extract more relevant information, refining our ability to filter false positives or negatives and, ultimately, to produce more defined new leads for prospective therapeutic interventions.

Acknowledgements

Rico Barsacchi and Varadharajan Sundaramurthy equally contributed to the manuscript.

References

Appelmelk, B.J., van Die, I., van Vliet, S.J., Vandenbroucke-Grauls, C.M., Geijtenbeek, T.B., and van Kooyk, Y. (2003). Cutting edge: carbohydrate profiling identifies new pathogens that interact with dendritic cell-specific ICAM-3-grabbing nonintegrin on dendritic cells. J. Immunol. *170*, 1635–1639.

Bach, H., Papavinasasundaram, K.G., Wong, D., Hmama, Z., and Av-Gay, Y. (2008). *Mycobacterium tuberculosis* virulence is mediated by PtpA dephosphorylation of human vacuolar protein sorting 33B. Cell Host Microbe 3, 316–322.

Banaiee, N., Kincaid, E.Z., Buchwald, U., Jacobs, W.R., Jr., and Ernst, J.D. (2006). Potent inhibition of macrophage responses to IFN-gamma by live virulent *Mycobacterium tuberculosis* is independent of mature mycobacterial lipoproteins but dependent on TLR2. J. Immunol. *176*, 3019–3027.

Beaulieu, A.M., Rath, P., Imhof, M., Siddall, M.E., Roberts, J., Schnappinger, D., and Nathan, C.F. (2010). Genome-wide screen for *Mycobacterium tuberculosis* genes that regulate host immunity. PLoS One *5*, e15120.

Benson, S.A., and Ernst, J.D. (2009). TLR2-dependent inhibition of macrophage responses to IFN-gamma is mediated by distinct, gene-specific mechanisms. PLoS One *4*, e6329.

Bongfen, S.E., Laroque, A., Berghout, J., and Gros, P. (2009). Genetic and genomic analyses of host–pathogen interactions in malaria. Trends Parasitol. *25*, 417–422.

Boucrot, E., Henry, T., Borg, J.P., Gorvel, J.P., and Meresse, S. (2005). The intracellular fate of *Salmonella* depends on the recruitment of kinesin. Science *308*, 1174–1178.

Brass, A.L., Huang, I.C., Benita, Y., John, S.P., Krishnan, M.N., Feeley, E.M., Ryan, B.J., Weyer, J.L., van der Weyden, L., Fikrig, E., *et al.* (2009). The IFITM proteins mediate cellular resistance to influenza A H1N1 virus, West Nile virus, and dengue virus. Cell *139*, 1243–1254.

Brennan, P.J., and Nikaido, H. (1995). The envelope of mycobacteria. Annu. Rev. Biochem. *64*, 29–63.

Brumell, J.H., and Grinstein, S. (2004). *Salmonella* redirects phagosomal maturation. Curr. Opin. Microbiol. *7*, 78–84.

Cappellini, M.D., and Fiorelli, G. (2008). Glucose-6-phosphate dehydrogenase deficiency. Lancet *371*, 64–74.

Collinet, C., Stoter, M., Bradshaw, C.R., Samusik, N., Rink, J.C., Kenski, D., Habermann, B., Buchholz, F., Henschel, R., Mueller, M.S., *et al.* (2010). Systems survey of endocytosis by multiparametric image analysis. Nature *464*, 243–249.

Desvignes, L., and Ernst, J.D. (2009). Interferon-gamma-responsive nonhematopoietic cells regulate the immune response to *Mycobacterium tuberculosis*. Immunity *31*, 974–985.

Diedrich, C.R., and Flynn, J.L. (2011). HIV-1/*Mycobacterium tuberculosis* coinfection immunology: how does HIV-1 exacerbate tuberculosis? Infect. Immun. *79*, 1407–1417.

Dyer, M.D., Murali, T.M., and Sobral, B.W. (2008). The landscape of human proteins interacting with viruses and other pathogens. PLoS Pathog. *4*, e32.

Epiphanio, S., Mikolajczak, S.A., Goncalves, L.A., Pamplona, A., Portugal, S., Albuquerque, S., Goldberg, M., Rebelo, S., Anderson, D.G., Akinc, A., *et al.* (2008). Heme oxygenase-1 is an anti-inflammatory host factor that promotes murine *Plasmodium* liver infection. Cell Host Microbe 3, 331–338.

Fortune, S.M., Solache, A., Jaeger, A., Hill, P.J., Belisle, J.T., Bloom, B.R., Rubin, E.J., and Ernst, J.D. (2004). *Mycobacterium tuberculosis* inhibits macrophage responses to IFN-gamma through myeloid differentiation factor 88-dependent and -independent mechanisms. J. Immunol. *172*, 6272–6280.

Gabriel, G., Herwig, A., and Klenk, H.D. (2008). Interaction of polymerase subunit PB2 and NP with importin alpha1 is a determinant of host range of influenza A virus. PLoS Pathog. *4*, e11.

Gamo, F.J., Sanz, L.M., Vidal, J., de Cozar, C., Alvarez, E., Lavandera, J.L., Vanderwall, D.E., Green, D.V., Kumar, V., Hasan, S., *et al.* (2010). Thousands of chemical starting points for antimalarial lead identification. Nature *465*, 305–310.

Guiguemde, W.A., Shelat, A.A., Bouck, D., Duffy, S., Crowther, G.J., Davis, P.H., Smithson, D.C., Connelly, M., Clark, J., Zhu, F., *et al.* (2010). Chemical genetics of *Plasmodium falciparum*. Nature *465*, 311–315.

Guiney, D.G., and Lesnick, M. (2005). Targeting of the actin cytoskeleton during infection by *Salmonella* strains. Clin. Immunol. *114*, 248–255.

Gutierrez, M.G., Master, S.S., Singh, S.B., Taylor, G.A., Colombo, M.I., and Deretic, V. (2004). Autophagy is a defense mechanism inhibiting BCG and *Mycobacterium tuberculosis* survival in infected macrophages. Cell *119*, 753–766.

Hao, L., Sakurai, A., Watanabe, T., Sorensen, E., Nidom, C.A., Newton, M.A., Ahlquist, P., and Kawaoka, Y. (2008). *Drosophila* RNAi screen identifies host genes important for influenza virus replication. Nature *454*, 890–893.

Hardt, W.D., Chen, L.M., Schuebel, K.E., Bustelo, X.R., and Galan, J.E. (1998). *S. typhimurium* encodes an activator of Rho GTPases that induces membrane ruffling and nuclear responses in host cells. Cell *93*, 815–826.

Harrison, R.E., Bucci, C., Vieira, O.V., Schroer, T.A., and Grinstein, S. (2003). Phagosomes fuse with late endosomes and/or lysosomes by extension of membrane protrusions along microtubules: role of Rab7 and RILP. Mol. Cell Biol. *23*, 6494–6506.

Holden, D.W. (2002). Trafficking of the *Salmonella* vacuole in macrophages. Traffic *3*, 161–169.

Jassal, M.S., and Bishai, W.R. (2010). Epidemiology and challenges to the elimination of global tuberculosis. Clin. Infect. Dis. *50 (Suppl. 3)*, S156–164.

Jayachandran, R., Sundaramurthy, V., Combaluzier, B., Mueller, P., Korf, H., Huygen, K., Miyazaki, T., Albrecht, I., Massner, J., and Pieters, J. (2007). Survival of mycobacteria in macrophages is mediated by coronin 1-dependent activation of calcineurin. Cell *130*, 37–50.

Jayachandran, R., Gatfield, J., Massner, J., Albrecht, I., Zanolari, B., and Pieters, J. (2008). RNA interference in J774 macrophages reveals a role for coronin 1 in mycobacterial trafficking but not in actin-dependent processes. Mol. Biol. Cell *19*, 1241–1251.

Jayaswal, S., Kamal, M.A., Dua, R., Gupta, S., Majumdar, T., Das, G., Kumar, D., and Rao, K.V. (2010). Identification of host-dependent survival factors for intracellular *Mycobacterium tuberculosis* through an siRNA screen. PLoS Pathog. *6*, e1000839.

Karlas, A., Machuy, N., Shin, Y., Pleissner, K.P., Artarini, A., Heuer, D., Becker, D., Khalil, H., Ogilvie, L.A., Hess, S., *et al.* (2010). Genome-wide RNAi screen identifies human host factors crucial for influenza virus replication. Nature *463*, 818–822.

Kaufmann, S.H. (2001). How can immunology contribute to the control of tuberculosis? Nat. Rev. Immunol. *1*, 20–30.

Kaufmann, S.H., and Parida, S.K. (2008). Tuberculosis in Africa: learning from pathogenesis for biomarker identification. Cell Host Microbe *4*, 219–228.

Kincaid, E.Z., and Ernst, J.D. (2003). *Mycobacterium tuberculosis* exerts gene-selective inhibition of transcriptional responses to IFN-gamma without inhibiting STAT1 function. J. Immunol. *171*, 2042–2049.

Konig, R., Stertz, S., Zhou, Y., Inoue, A., Hoffmann, H.H., Bhattacharyya, S., Alamares, J.G., Tscherne, D.M., Ortigoza, M.B., Liang, Y., *et al.* (2010). Human host factors required for influenza virus replication. Nature *463*, 813–817.

Kuijl, C., Savage, N.D., Marsman, M., Tuin, A.W., Janssen, L., Egan, D.A., Ketema, M., van den Nieuwendijk, R., van den Eeden, S.J., Geluk, A., *et al.* (2007). Intracellular bacterial growth is controlled by a kinase network around PKB/AKT1. Nature *450*, 725–730.

Kumar, D., Nath, L., Kamal, M.A., Varshney, A., Jain, A., Singh, S., and Rao, K.V. (2010). Genome-wide analysis of the host intracellular network that regulates survival of *Mycobacterium tuberculosis*. Cell *140*, 731–743.

Kyei, G.B., Vergne, I., Chua, J., Roberts, E., Harris, J., Junutula, J.R., and Deretic, V. (2006). Rab14 is critical for maintenance of *Mycobacterium tuberculosis* phagosome maturation arrest. EMBO J. *25*, 5250–5259.

Langhi, D.M., Jr., and Bordin, J.O. (2006). Duffy blood group and malaria. Hematology *11*, 389–398.

Lenard, J. (1978). Virus envelopes and plasma membranes. Annu. Rev. Biophys. Bioeng. *7*, 139–165.

Maier, A.G., Duraisingh, M.T., Reeder, J.C., Patel, S.S., Kazura, J.W., Zimmerman, P.A., and Cowman, A.F. (2003). *Plasmodium falciparum* erythrocyte invasion through glycophorin C and selection for Gerbich negativity in human populations. Nat. Med. *9*, 87–92.

Marsman, M., Jordens, I., Kuijl, C., Janssen, L., and Neefjes, J. (2004). Dynein-mediated vesicle transport controls intracellular *Salmonella* replication. Mol. Biol. Cell *15*, 2954–2964.

Matrosovich, M., Stech, J., and Klenk, H.D. (2009). Influ.enza receptors, polymerase and host range. Rev. Sci. Technol. *28*, 203–217.

Meresse, S., Steele-Mortimer, O., Moreno, E., Desjardins, M., Finlay, B., and Gorvel, J.P. (1999). Controlling the maturation of pathogen-containing vacuoles: a matter of life and death. Nat. Cell Biol. *1*, E183–188.

Nau, G.J., Richmond, J.F., Schlesinger, A., Jennings, E.G. Lander, E.S., and Young, R.A. (2002). Human macrophage activation programs induced by bacterial pathogens. Proc. Natl. Acad. Sci. U.S.A. *99*, 1503–1508.

Noss, E.H., Pai, R.K., Sellati, T.J., Radolf, J.D., Belisle, J., Golenbock, D.T., Boom, W.H., and Harding, C.V. (2001). Toll-like receptor 2-dependent inhibition of macrophage class II MHC expression and antigen processing by 19-kDa lipoprotein of *Mycobacterium tuberculosis*. J. Immunol. *167*, 910–918.

Ottenhoff, T.H., Verreck, F.A., Lichtenauer-Kaligis, E.G. Hoeve, M.A., Sanal, O., and van Dissel, J.T. (2002). Genetics, cytokines and human infectious disease: lessons from weakly pathogenic mycobacteria and salmonellae. Nat. Genet. *32*, 97–105.

Pethe, K., Sequeira, P.C., Agarwalla, S., Rhee, K., Kuhen, K., Phong, W.Y., Patel, V., Beer, D., Walker, J.R., Duraiswamy, J., *et al.* (2010). A chemical genetic screen in *Mycobacterium tuberculosis* identifies carbon-source-dependent growth inhibitors devoid of *in vivo* efficacy. Nat. Commun. *1*, 57.

Prudencio, M., Rodrigues, C.D., Hannus, M., Martin, C., Real, E., Goncalves, L.A., Carret, C., Dorkin, R., Rohl, I., Jahn-Hoffmann, K., *et al.* (2008). Kinome-wide RNAi screen implicates at least 5 host hepatocyte kinases in *Plasmodium sporozoite* infection. PLoS Pathog. *4*, e1000201.

Rodrigues, C.D., Hannus, M., Prudencio, M., Martin, C., Goncalves, L.A., Portugal, S., Epiphanio, S., Akinc, A., Hadwiger, P., Jahn-Hofmann, K., *et al.* (2008). Host scavenger receptor SR-BI plays a dual role in the establishment of malaria parasite liver infection. Cell Host Microbe *4*, 271–282.

Saleh, M.T., and Belisle, J.T. (2000). Secretion of an acid phosphatase (SapM) by *Mycobacterium tuberculosis* that is similar to eukaryotic acid phosphatases. J. Bacteriol. *182*, 6850–6853.

Santos, R.L., Zhang, S., Tsolis, R.M., Kingsley, R.A., Adams, L.G., and Baumler, A.J. (2001). Animal models of *Salmonella* infections: enteritis versus typhoid fever. Microbes Infect. *3*, 1335–1344.

Shwegmann, A., and Brombacher, F. (2008). Host-directed drug targeting of factors hijacked by pathogens. Sci. Signal. *1*, re8.

Shapira, S.D., Gat-Viks, I., Shum, B.O., Dricot, A., de Grace, M.M., Wu, L., Gupta, P.B., Hao, T., Silver, S.J., Root, D.E., *et al.* (2009). A physical and regulatory map of host-influenza interactions reveals pathways in H1N1 infection. Cell *139*, 1255–1267.

Snijder, B., Sacher, R., Ramo, P., Damm, E.M., Liberali, P., and Pelkmans, L. (2009). Population context determines cell-to-cell variability in endocytosis and virus infection. Nature *461*, 520–523.

Stein, M.A., Leung, K.Y., Zwick, M., Garcia-del Portillo, F., and Finlay, B.B. (1996). Identification of a *Salmonella* virulence gene required for formation of filamentous structures containing lysosomal membrane glycoproteins within epithelial cells. Mol. Microbiol. *20*, 151–164.

Stertz, S., and Shaw, M.L. (2011). Uncovering the global host cell requirements for influenza virus replication via RNAi screening. Microbes Infect. *13*, 516–525.

Sui, B., Bamba, D., Weng, K., Ung, H., Chang, S., Van Dyke, J., Goldblatt, M., Duan, R., Kinch, M.S., and Li, W.B. (2009). The use of random homozygous gene perturbation to identify novel host-oriented targets for influenza. Virology *387*, 473–481.

Taubenberger, J.K., and Kash, J.C. (2010). Influenza virus evolution, host adaptation, and pandemic formation. Cell Host Microbe *7*, 440–451.

Ting, L.M., Kim, A.C., Cattamanchi, A., and Ernst, J.D. (1999). *Mycobacterium tuberculosis* inhibits IFN-gamma transcriptional responses without inhibiting activation of STAT1. J. Immunol. *163*, 3898–3906.

Uchiya, K., Barbieri, M.A., Funato, K., Shah, A.H., Stahl, P.D., and Groisman, E.A. (1999). A *Salmonella* virulence protein that inhibits cellular trafficking. EMBO J. *18*, 3924–3933.

Vazquez-Torres, A., Xu, Y., Jones-Carson, J., Holden, D.W., Lucia, S.M., Dinauer, M.C., Mastroeni, P., and Fang, F.C. (2000). *Salmonella* pathogenicity island 2-dependent evasion of the phagocyte NADPH oxidase. Science *287*, 1655–1658.

Vergne, I., Chua, J., and Deretic, V. (2003). Tuberculosis toxin blocking phagosome maturation inhibits a novel Ca2+/calmodulin-PI3K hVPS34 cascade. J. Exp. Med. *198*, 653–659.

Vergne, I., Chua, J., Singh, S.B., and Deretic, V. (2004). Cell biology of *Mycobacterium tuberculosis* phagosome. Annu. Rev. Cell. Dev. Biol. *20*, 367–394.

Volkman, H.E., Pozos, T.C., Zheng, J., Davis, J.M., Rawls, J.F., and Ramakrishnan, L. (2010). Tuberculous granuloma induction via interaction of a bacterial secreted protein with host epithelium. Science *327*, 466–469.

Walburger, A., Koul, A., Ferrari, G., Nguyen, L., Prescianotto-Baschong, C., Huygen, K., Klebl, B., Thompson, C., Bacher, G., and Pieters, J. (2004). Protein kinase G from pathogenic mycobacteria promotes survival within macrophages. Science *304*, 1800–1804.

Watanabe, T., Watanabe, S., and Kawaoka, Y. (2010). Cellular networks involved in the influenza virus life cycle. Cell Host Microbe *7*, 427–439.

Weatherall, D.J. (2008). Genetic variation and susceptibility to infection: the red cell and malaria. Br. J. Haematol. *141*, 276–286.

How One Cell Eats Another: Principles of Phagocytosis

6

Sylvain Tollis, Navin Gopaldass, Thierry Soldati and Robert G. Endres

Abstract

Phagocytosis is the fundamental cellular process by which eukaryotic cells bind and engulf particles by deforming their plasma membrane. Particle engulfment involves particle recognition by cell-surface receptors, signalling, and remodelling of the actin cytoskeleton to guide the membrane around the particle in a zipper-like fashion. The signalling complexity is daunting, involving hundreds of different molecular species during the initial stages of engulfment. For instance, the well-characterized immune Fcγ and the integrin CR3 receptors signal to tyrosine kinases and Rho GTPases, ultimately regulating a wide variety of proteins, which direct actin polymerization and myosin-motor proteins for force generation and contraction. Despite the signalling complexity, phagocytosis also depends strongly on simple biophysical parameters, such as particle shape and cell stiffness, or membrane biophysical properties that are independent of the type of cell or particle. We argue that these emergent, universal features are particularly important to address in order to explain this evolutionary well-conserved and robust mechanism. In this chapter we review these universal features to describe the principles of phagocytosis. Specifically, we use a recently published model of phagocytic engulfment as a guide. Finally, we discuss state-of-the-art live-cell fluorescence microscopy, recently used to elucidate the dynamics of phospholipids, actin polymerization and myosins in the particle shape recognition by the amoeba *Dictyostelium*.

Introduction

Phagocytosis is the ancient, evolutionarily conserved process by which eukaryotic cells bind, engulf, and destroy particles and cells larger than 0.5 μM in diameter (Underhill and Ozinsky, 2002; Swanson and Hoppe, 2004; Swanson, 2008; Boulais *et al.*, 2010). The importance of phagocytosis is derived from its two main functions: (1) a feeding mechanism in single-cell organisms such as the *Dictyostelium* amoeba (Bretschneider *et al.*, 2009), and (2) clearance of pathogens, apoptotic and senescent cells from the vertebrate body by immune cells (Greenberg and Grinstein, 2002; Conner and Schmid, 2008). As part of our immune defence, phagocytosis is mainly performed by professional phagocytes, including macrophages, neutrophils and dendritic cells. Initiation of phagocytosis occurs with recognition of the target particle either directly or via an opsonising molecule. For instance, the Fc portion of immunoglobulin G (IgG) is recognized by the cell-surface receptor FcγRIIa

(Daëron, 1997; Ravetch and Bolland, 2001; Dupuy and Caron, 2008; Groves *et al.*, 2008). Ligand–receptor binding triggers intracellular signalling (Goldstein *et al.*, 2001; Underhill and Ozinsky, 2002; Groves *et al.*, 2008), resulting in remodelling of the actin cytoskeleton (Castellano *et al.*, 2001; May and Machesky, 2001) and coherent growth of cell membrane around the particle to form the phagocytic cup (Swanson and Hoppe, 2004; Swanson, 2008). Eventually, the leading edge of the growing cup closes, and a membrane vesicle enclosing the particle (phagosome) is pulled into the cell. Subsequently, the phagosome fuses with vesicles delivering the acidifying V-ATPase proton pump (Kinchen and Ravichandran, 2008) and digestive enzymes (Silva *et al.*, 2007; Yu *et al.*, 2008), as well as oxygen radical-producing enzymes (Hampton *et al.*, 1998; Segal, 2005) to destroy the particle.

Phagocytosis is a highly conserved cellular mechanism. A recent exhaustive study compared the proteome of latex beads-containing phagosomes isolated from murine macrophages, *Drosophila* S2 cells and *Dictyostelium* amoeba (Boulais *et al.*, 2010). This comparison revealed that most of the proteins involved in uptake and maturation of the phagosome are part of a very conserved 'ancient core' of phagosomal proteins. This includes proteins involved in the cytoskeleton, in membrane trafficking, in acidification and digestion, as well as a large number of small GTPases involved in signalling. This confirms at a proteomic level what earlier studies have shown about the conservation of the mechanisms, dynamics and morphology of the endocytic and phagocytic pathways (Neuhaus and Soldati, 1999, 2002).

The biochemical pathways involved in phagocytosis are complex. During the initial phase when the particle attaches to the phagocyte, multiple signals are simultaneously activated. In parallel, signals can trigger cytoskeletal alteration, modifications in membrane trafficking, production of inflammatory responses through cytokines and chemokines, and activation of killing mechanisms to digest the particle. The complexity originates in part from the fact that dozens of cell-surface receptors can contribute to the recognition of a large variety of ligand molecules and subsequent particle engulfment (Underhill and Ozinsky, 2002; Kraft and Kinet, 2007; Groves *et al.*, 2008). On top of that, downstream signals originating from the receptors can interact, either cooperatively or destructively. Furthermore, the particle itself, if a bacterium, may try to escape or enhance internalization. In addition to the Fcγ receptor (FcγR), the complement receptor 3 (CR3) of the integrin receptor family (Tohyama and Yamamura, 2008) is another widely studied receptor-type involved in phagocytosis.

Professional phagocytes express different classes of Fc receptors. Specifically, the ITAM domain of Fc receptors is phosphorylated upon ligand–receptor binding which leads to the creation of a signalling complex and recruitment of signalling molecules such as kinases (e.g. Syk) and GTPases (e.g. Cdc42, Rac). This activates a chain of chemical reactions to promote engulfment. In contrast, receptors that contain an ITIM domain on their intracellular portion will recruit phosphatases, which inhibit signalling. A complex interplay between the excitatory and inhibitory pathways regulates particle engulfment. Unlike FcR-mediated phagocytosis, CR3-mediated phagocytosis requires a second activation step (after binding to the iC3b ligand) to trigger the onset of phagocytosis (Berget *et al.*, 1984; Sengelov *et al.*, 1993; Jones *et al.*, 1998; Blystone and Brown, 1999). The second activation signal, although not completely understood, is likely to involve activation of phosphokinase C (PKC) (Wright and Griffin, 1985; Aderem and Underhill, 1999), and Rap1, a small GTPase of the Ras family (Caron *et al.*, 2000). Complement receptor and Fc receptor coligation can produce cooperative effects, and particles coated with sufficient IgG for FcR activation are engulfed eventually, if they are also coated with complement molecules (Ehlenberger and Nussenzweig, 1977). In these experiments, ligation of FcR produces the second activation signal required by the complement receptor machinery.

There is a key difference in phagocytic cup shape between FcR- and CR3-mediated phagocytosis. FcγR-mediated phagocytosis proceeds through membrane protrusions and leads to thin cups (Niedergang and Chavrier, 2003; Champion and Mitragotri, 2006), whereas in CR3-mediated phagocytosis, particles appear to sink into the cell (Allen and Aderem, 1996; Le Cabec *et al.*, 2002). Efficient internalization involves many regulators, and is beyond the scope of this chapter. However, four of them are important for signal integration, the inflammatory response, and particle degradation, making them important ingredients for modelling. These key players are phosphoinositide 3 kinase (PI3K), phospholipase C (PLC), the class of Rho GTPases, and PKC.

The plasma membrane is a reactive surface and its regulation is key for successful phagocytic engulfment and membrane trafficking (Fig. 6.1). Phospholipids of the phosphoinositide family are important second messengers with catalytic activities (Grinstein, 2010). PI3K catalyses the phosphorylation and conversion of phospholipid PI(4,5)P_2 to PI(3,4,5)P_3 (Bohdanowicz *et al.*, 2010). Despite their importance, these two comprise only about 1% of the phospholipids in the cytoplasmic leaflet of the plasma membrane. These lipids serve as docking sites for a large variety of cytosolic proteins that are recruited to the cytosolic leaflet of the plasma membrane and cytoplasmic organelles via the binding of domains such as pleckstrin homology (PH) domains. Binding of these proteins to specific phosphoinositide isoforms promote their recruitment and subsequent activation of signalling cascades. For instance, PI(3,4,5)P_3 allows the recruitment of myosins to the phagocytic cup whose contractile activity is crucial for engulfment (Araki *et al.*, 2003). PI3K is also involved in membrane fusion needed for the completion of engulfment (Cox *et al.*, 1999; Swanson *et al.*, 1999). Inhibition of PI3K does not alter initial attachment of the particle and signalling. However, it blocks the later stages of engulfment but the mechanism is still debated (Araki *et al.*, 1996; Cox *et al.*, 1999; Gold *et al.*, 1999). Furthermore, inhibition of PI3K blocks the activation of Arf1 and Rac2, as well as the deactivation of Arf6 and Cdc42. This indicates a PI(3,4,5)P_3-dependent transition in the signalling downstream of Fc receptors. PLC is

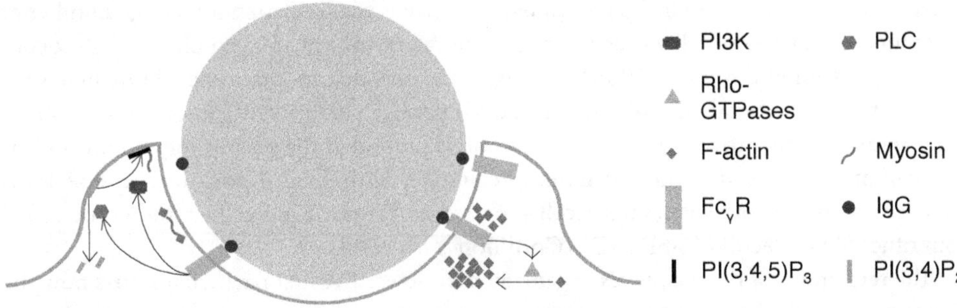

Figure 6.1 Schematic of Fc-receptor mediated phagocytosis. Fc receptor – IgG binding triggers Rho-GTPases recruitment ultimately leading to actin polymerization (right), but also impact phospholipids signalling via PI3K recruitment (left), which in turns promotes myosin motors conveying at the phagocytic cup.

an enzyme, which activates PKC. Different isoforms of the enzyme are involved in actin remodelling, cytokines production and activation of killing mechanisms. Furthermore, PLC catalyses the hydrolysis of PI(4,5)P_2 into two important second messenger molecules (Hao *et al.*, 2009), which in turn drive various cell responses such as proliferation, apoptosis, or cytoskeleton remodelling.

Clustering of FcγR upon binding to IgG recruits small GTPases of the Rho family [typically, Rho, Rac and Cdc42 (Caron and Hall, 1998)], which, via proteins of the WASP family, activate the Arp2/3 complex (Underhill and Ozinsky, 2002; Swanson and Hoppe, 2004). The latter promotes branching of actin filaments, leading both to an increase in the number of uncapped ends and to an isotropic growth of the actin network (Medalia *et al.*, 2007; Groves *et al.*, 2008). In the case of complement-mediated phagocytosis, only Rho (Caron and Hall, 1998), which induces the assembly of contractile actomyosin filaments, is involved (and not Rac or Cdc42). Note that these two distinctive pathways initiated by activation of different receptors involve both the actin machinery (Flannagan *et al.*, 2010). Additionally, the phagocytic cup is enriched in gelsolin (Yin *et al.*, 1981; Serrander *et al.*, 2000; Arora *et al.*, 2005), coronin (Bretschneider *et al.*, 2009), and other regulators of actin polymerization. Gelsolin is a lipid-bound protein, which has a high affinity for actin via an actin-binding site, and severs actin filaments. Additionally it shows a high affinity with PIP phospholipids. Interestingly, gelsolin is inactivated when in contact with PI(4,5)P_2. This underlines the strong interaction between membrane trafficking and cytosolic actin remodelling during phagocytosis. Coronin prevents actin branching by removing the Arp2/3 complex from branched actin structures and therefore preventing further polymerization (Cai *et al.*, 2008).

To study phagocytosis in a single-cell organism *Dictyostelium discoideum* is an excellent model system, which historically led to the discovery of new proteins involved in this process such as coronin. In addition, upon starvation, *Dictyostelium* behaves as a 'facultative metazoan', as it undergoes a differentiation cycle during which about 100,000 cells come together to form a slug. This slug is a true multicellular organism driven by self–non-self recognition pathways (Benabentos *et al.*, 2009) and is composed of at least five different cell types, including a population of Sentinel cells that patrol the slug for xenobiotics and foreign particles (Chen *et al.*, 2007). The slug then further differentiates in a fruiting body composed of a stalk supporting a mass of spores (reviewed in Mahaedo and Parent, 2006). Although the molecular machinery for particle/bacteria recognition and binding has not yet been completely identified in *Dictyostelium*, evidence is accumulating for a large number of receptors for particle uptake, specific for a wide variety of ligands, proteins and glycoconjugates (Cosson and Soldati, 2008). For example, an integrin-like molecule named SibA (similar to integrin beta) (Cornillon *et al.*, 2006) has been identified. This protein has structural features similar to integrin beta chains and is present at the plasma membrane and, as for mammalian integrins, SibA co-imunoprecipitates with Talin. Knock-out of SibA leads to drastic reduction of phagocytosis, cell adhesion and spreading. Further studies identified four other Sib molecules (SibB to E) (Cornillon *et al.*, 2008).

The remainder of the chapter is organized as follows. The first part will discuss how the concentration and distribution of ligands on the particle surface strongly influences whether or not phagocytic engulfment is completed. In the 1970s this gave rise to the idea that engulfment is a progressive, zipper-like mechanism that requires sequential ligand–receptor binding. We will review a recently introduced three-dimensional, stochastic biophysical model of the zipper mechanism. This model, though minimal, allows computer simulation of the engulfment of particles of various sizes and shapes, including spiral and rod-shaped particles reminiscent of bacteria. Highly curved shapes are not taken up, in line with recent experimental results. This model and associated experiments predict that, surprisingly, engulfment proceeds even without actin-driven forces in a large regime of parameter values, albeit more slowly and with highly variable phagocytic cups. In the light of recent experiments on the later stages of engulfment, the question of bistability and cell decision-making during engulfment will be addressed. The second half of the chapter will be devoted to fluorescence live-cell imaging of phagocytosis, which provides important information on the localization of biological molecules and their time dependence. Specifically, we will investigate the role of actin waves and myosins in the recognition of yeast particles by phagocytic *Dictyostelium* cells. Finally, we will examine other fields of biology in which similar ideas and methods may successfully be applied. Indeed, zipper-like processes are encountered in bacterial sporulation and fruit fly embryo dorsal closure. Finally, we discuss how the influence of particle-shape on phagocytosis highlights the need for a systematic investigation of host–pathogen interactions with potential application in improved design of particles as vehicles for drug delivery.

Physical and geometric principles of phagocytosis

The relevance of physical sciences in the study of phagocytosis originates from the fact that engulfment requires the deformation of the cell's plasma membrane. This deformation involves the actin cytoskeleton and molecular motor proteins, which follow the laws of polymer physics, statistical mechanics and thermodynamics. Specifically, spreading of the cell membrane over the particle involves actin-driven cell-shape changes similar to the processes involved in cell migration and adhesion (Svitkina and Borisy, 1999; Charras *et al.*, 2006; Dupuy and Caron, 2008; Chan *et al.*, 2009; Colombelli *et al.*, 2009). In addition to the cell's requirements, there are also a number of mechanical and geometrical restrictions on the particle. Interestingly, completion of phagocytic uptake depends on the shape and initial orientation of the particle in relation to the cell surface (Champion and Mitragotri, 2006, 2009). For instance, experiments demonstrate that elongated spheroid polystyrene particles coated with IgG are more efficiently engulfed when presented to the phagocyte with their tip first. Furthermore, a recent study demonstrates that the biophysical requirements of phagocytosis lead to either complete phagocytosis or stalled cups due to the presence of a mechanical bottleneck (Van Zon *et al.*, 2009). Interestingly, the same study showed that engulfment even appears to proceed in cells treated with modest amounts of cytochalasin D, an inhibitor of actin polymerization, indicating that biochemical pathways may not always be necessary for the initial stages of phagocytosis. The softness of the particle also has important effects on the completion of phagocytosis. It was reported recently that macrophages offered particles of identical chemical properties but different rigidity show a strong preference for engulfing the rigid particles. Engulfment of soft particles can be restored by injecting active Rac1 in the macrophage, indicating the presence of Rac1-dependent mechanisms for sensing the mechanical properties of particles during phagocytosis (Beningo and Wang, 2002).

Recent experiments provide new insights into the biophysical mechanism for driving the membrane around the particle, suggesting a ratchet-type mechanism. Once started, phagocytosis progresses unidirectionally and irreversibly (Herant *et al.*, 2005). This irreversible membrane progression is further supported by the loss of lipid and protein mobility at the phagocytic cup, observed using fluorescence recovery after photobleaching (FRAP) (Corbett-Nelson *et al.*, 2006). Several models have proposed actin polymerization as a mechanism of force generation (see (Ananthakrishnan and Ehrlicher, 2007) and references therein), but recent experiments based on fluorescent speckle microscopy demonstrate that actin does not directly push the membrane outwards. Instead, by filling gaps provided by membrane fluctuations (or other types of membrane movement), actin polymerization acts like a ratchet preventing the membrane from moving backwards (Herant *et al.*, 2006). The relevance of such 'Brownian' ratchets in biology has previously been emphasized (Mogilner *et al.*, 1996, 2003). A recent model, reviewed in the following sections, shows that a ratchet mechanism, together with energetic restrictions in membrane bending and stretching, can not only spontaneously lead to phagocytic uptake but also account for the influence of shape on phagocytosis and other biophysical aspects of engulfment.

Ligand density and zipper mechanism

Attempts were made more than 30 years ago to understand the physical mechanisms governing membrane deformations during phagocytosis. In the 1970s, Griffin and collaborators demonstrated that sequential ligand–receptor binding throughout engulfment was required for complete uptake of the particle (Griffin *et al.*, 1975, 1976). Specifically, these authors studied the uptake of sheep erythrocytes opsonized with immunoglobulins IgG or complement ligands by mouse peritoneal macrophages. Using an initial incubation of cells at 4°C, a temperature at which erythrocytes attach to macrophages but are not taken up, followed by a heating up to 37°C to trigger engulfment, these authors managed to separate the initial attachment phase from the engulfment phase of phagocytosis. After initial attachment, they reported that the *phagocytic index* (fraction of engulfed particles) was considerably reduced, either by neutralizing free-membrane Fc receptors with antimacrophage IgG in one experiment, or by altering unbound complement ligand using a trypsin treatment. This indicated that ligand and receptor molecules not involved in the initial attachment are crucial for the completion of engulfment. In other words, phagocytic uptake occurs via successive zipper-like ligand–receptor binding (Fig. 6.2), and not by an all-or-none mechanism triggered at the onset of phagocytosis. Indeed, such a mechanism would require that some biochemical

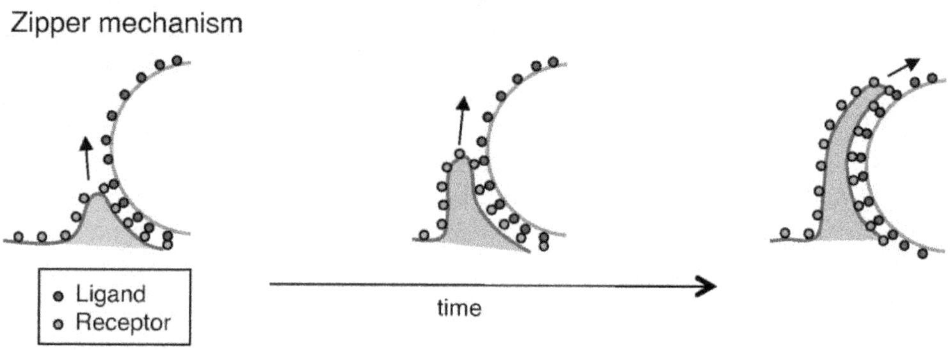

Figure 6.2 Schematic of the zipper mechanism. Cell-membrane receptors (empty dots) and ligands on the particle surface (filled dots) are sequentially engaged in bond formation, resulting in progression of engulfment with time (from left to right).

signal, generated by ligand–receptor binding from the initial attachment of the particle, exceeds a threshold and triggers the phagocytic response. In this case, the amplitude of the response would be independent on the signal magnitude, as long as it is beyond the threshold. Such a mechanism would indicate that, soon after attachment, the cell makes the decision whether to engulf the particle or not. In the zipper mechanism as envisaged by Griffin and Silverstein, engulfment progresses as long as membrane receptors find a significant amount of ligand on the particle to bind to until engulfment is completed, or stop in the absence of sufficient local molecular resources.

In both of these experiments it was still possible that the initial ligand–receptor bounds are altered by the treatment, and that the number of bonds fall below a threshold necessary to trigger the ingestion of erythrocytes. To exclude this eventuality and to validate the zipper mechanism, the same authors performed a second round of experiments, in which mouse peritoneal macrophages were presented with mesenteric lymph-node lymphocytes as particles to ingest. Lymphocytes were mixed with fluorescent anti-IgG at 4°C, and showed a roughly uniform distribution of fluorochrome under the microscope. Some lymphocytes were incubated another 30 minutes at 25°C, which leads to a capping of anti-IgG, and concentration at one pole of the lymphocyte. Consequently, particles of the first group were entirely coated with ligands, whereas particles of the second group had the same amount of ligands concentrated on one half of their surface. As a result of these experiments, particles entirely coated with ligands were taken up by macrophages whereas particles coated only on one side were not ingested. Furthermore, the subsequent treatment of half-coated particles with immunoglobulin restored the uniform coating, and partially engulfed particles were then completely ingested. There results confirmed that the formation of new ligand–receptor bounds throughout engulfment was required for efficient phagocytosis, and therefore established the zipper as the driving mechanism for engulfment.

Several subsequent experiments supported the zipper mechanism. For instance, as phagocytosis is locally controlled in this mechanism, the phagocytic cup and eventual phagosomal membrane should be a very tight fit to the particle. Indeed several imaging experiments have showed this to be the case for a variety of phagocytes and particles [see Swanson and Hoppe (2004) and references therein, (Jeon and Jeon, 1976; Isberg and Van Nhieu, 1995; Gerisch *et al.*, 2009)]. Furthermore, macrophages that spread onto IgG-opsonized coverslips in a process called frustrated phagocytosis, create a very narrow space in the region of contact (Kaplan, 1977), much narrower than the spreading on non-opsonized coverslips. This shows that Fc-receptor binding to IgG ligand on the coverslip enhances the spreading to make a tighter fit in accordance with the zipper mechanism.

Other experiments report that the simple zipper mechanism is insufficient for understanding phagocytic engulfment. For instance, while CR3-mediated phagocytosis was shown to be zipper-like (Tohyama and Yamamua, 2008), it appears more like spreading on non-opsonized surfaces (Wright *et al.*, 1988; Hynes, 1992): less pseudopod extension, more irregular and wider phagocytic cups with high sensitivity to disruption of microtubules (Wright and Silverstein, 1982). Since the CR3 receptor belongs to the integrin family, the similarities with cell spreading on non-opsonized surfaces are not surprising. However, how can different receptor types lead to an overall similar physical mechanism, i.e. zipper, but with very different cup shapes? And how can the zipper mechanism lead to the particle shape and stiffness dependences? The simplistic zipper mechanism alone may not be able to address these issues.

Some types of engulfment seem to better fit the alternative trigger mechanism in which a cellular decision activates the complete engulfment after particle attachment. This is, for instance, the case for macropinocytosis. Under the influence of specific growth factors, macrophages start to produce large membrane ruffles (Racoosin and Swanson, 1989). While subsiding back onto the cell, some of these ruffles may enclose nearby particles, leading to their internalization. Unlike in zipper-mediated engulfment, the resulting phagosome is usually larger than the particle, and there is no close contact between the cell membrane and the particle surface. The ruffle is the cell's response to an increase in growth factor, and this response is controlled more globally by the amount of growth factor in the environment, independent of particle size and geometry. As an example, *Salmonella typhimurium* uses macropinocytosis to infect macrophages. Shortly after binding the bacterium, the cell shows a burst of ruffling activity on its entire surface, not only at the site of binding (Takeuchi, 1967; Francis *et al.*, 1992). As a side-effect of this ruffling activity, non-invasive bacteria (Galan and Hayman, 1992), or even latex particles (Francis *et al.*, 1993) may be internalized. Engulfment of *S. typhimurium* is therefore, to some extent, closer to a trigger mechanism than a zipper mechanism.

As a consequence, a complete picture of the biophysical mechanisms of phagocytic engulfment is likely to blend ingredients of both the zipper and the trigger mechanism. The timing and spatial localization of biochemical signals is likely to be the key to reconcile these apparently different mechanisms. For instance, the study of phagocytosis by *Amoeba proteus*, which creates a phagosome scaled to the particle size but almost without touching the particle (Jeon and Jeon, 1976), shows features of both the zipper and the trigger mechanism, and hence could provide new insight into answering these questions. A better understanding of the phagocytic engulfment process is also likely to be provided by the study of receptor dynamics during engulfment. Receptor dynamics appear to be controlled by the cell through actin polymerization and tyrosine phosphorylation (Andrews *et al.*, 2008). Specifically, phagocytes probe their environment via actin-rich protrusions in order to detect and attach poorly opsonized or highly mobile particles, which will provide more efficient particle attachment than simple receptor diffusion (Kress *et al.*, 2007; Jaumouille and Grinstein, 2010). Receptors dynamics and clustering are relevant throughout engulfment (Sobota *et al.*, 2005).

The zipper mechanism is the underlying assumption in a number of recent modelling works in phagocytosis (Herant *et al.*, 2006; Van Zon *et al.*, 2009) and endocytosis (Gao *et al.*, 2005; Decuzzi and Ferrari, 2007, 2008), mainly addressing the influence of cell membrane tension and ligand–receptor bond density on engulfment. Despite the general acceptance of the zipper mechanism, many of its biophysical requirements are insufficiently understood. Questions, so far unanswered, include what the energetic requirements are for the zipper mechanism, specifically what role actin polymerization plays in its progression during phagocytosis, and also whether the zipper mechanism can explain the particle-shape dependence of phagocytosis. These issues were addressed recently by using the first stochastic simulation model of phagocytic engulfment via the zipper mechanism (Tollis *et al.*, 2010). Thermal membrane fluctuations, assumed to play a major in phagocytosis (Ji *et al.*, 2008), provide the energy source to locally deform the membrane and to build further ligand–receptor bonds for zippering the membrane around the particle. Actin polymerization makes ligand–receptor bonds effectively *irreversible*, i.e. reinforced and stabilized for a significant amount of time. In other words, polymerizing actin fills the gaps left by the fluctuating membrane between the membrane itself and the underlying cortex. This remodelling of actin network is only possible close to the particle, where ligand–receptor signalling promotes the remodelling. Far from the particle, membrane fluctuations are reversible since they are not supported by actin polymerization. This ratchet, driven by actin polymerization to rectify disordered fluctuations of the membrane, allows the pumping of thermal energy into the system and its utilization. This energy gain is compared with the energy loss in deforming the membrane. Specifically, in this model the cell membrane and the actin cortex are described by a Helfrich-type energy function (Helfrich, 1973), including contributions from ligand–receptor binding, membrane bending and stretching, as well as a cell volume constraint. Membrane parameters are chosen to effectively describe the cell plasma membrane with its underlying actin cortex. The model was implemented using finite-temperature Monte Carlo simulations of the discretized cell membrane (see Tollis *et al.*, 2010, for details). The main results of this model are discussed in the subsections below.

Particle size, shape and stiffness

During an immune response, the main purpose of phagocytosis is to clear pathogenic bacteria from our bodies. Bacteria exist in a number of shapes and sizes and possess very different chemical properties. A striking example comes from the study of the enteropathogen *Campylobacter jejuni*. Dependent on the situation, the bacterium can be found in different shapes, including spirals, rods, cocci and doughnuts. Electron microscopic studies reveal that the shape of an individual *Campylobacter* bacterium in a colony differs depending on their location in the colony (Lai-king *et al.*, 1985). At the periphery, cells are almost all spirals, while in the centre of the colony cells are coccus shaped. Unusual doughnut-shaped cells are observed at the periphery as well. Furthermore, young or actively growing cells are mainly spiral shaped. Older cells undergo a degenerative change to coccoid forms. The doughnut-shape appears to be an intermediate stage between spirals and cocci. Why do cells in these colonies have different shapes? In the context of phagocytosis, a natural answer would be that bacteria try to escape engulfment. Therefore diverse shapes may minimize the likelihood of being ingested.

The question of why particle shape influences phagocytic engulfment was recently addressed by Champion and Mitragotri (2006). They studied phagocytosis by alveolar macrophages using polystyrene particles of various sizes and shapes, including elongated rod-like and spiral shapes reminiscent of actual bacteria (Champion and Mitragotri, 2006). They observed that as long as particles are smaller than the cell, their size is of little importance in the completion of engulfment. However, even if all shapes were able to initiate phagocytosis, spherical particles are taken up more easily than elongated particles and spiral-shaped particles. More interestingly, completion of engulfment depends on how the particle sits on the cell membrane. Elongated spheroid polystyrene particles coated with IgG are more efficiently engulfed when presented tip first to the phagocyte. The local shape at the point of attachment determines the complexity of the actin structure that must be created to initiate phagocytosis and allow the membrane to move over the particle. Failure to create the required actin structure results in simple membrane spreading and not internalization. These authors used the reported findings to drive the engineering of drug particles optimized for high-efficiency delivery to target cells or tissues. In a follow-up study they increased the delivery of drugs to target cells, by loading them into particles whose shape minimizes uptake by macrophages (Champion and Mitragotri, 2009).

Similar results on how particle shape and orientation are key in determining the outcome of phagocytosis were obtained with stochastic simulations (Tollis *et al.*, 2010). Specifically, engulfment of spherical particles with different radii ranging from 1.2 to 3.8 μm was simulated. Fig. 6.3 shows that engulfment progresses normally for both small and large particles. Hence the zipper mechanism is sufficiently robust to allow engulfment of differently sized particles using the same set of biophysical parameters (although engulfment of large particles requires more time). Large particles with radii greater than 2.5 μm were taken up via noticeably more regular phagocytic cups than were small particles with radii less than 1.5 μm. This indicates that the cell needs to save resources, like membrane area or curvature when engulfing large particles.

Not only geometrical, but also mechanical properties of the particle influence engulfment. In recent experiments Beningo and Wang (2002) prepared polyacrylamide microbeads of variable rigidity, coated with IgG. The stiffness of the beads was controlled by varying the concentration of the bis-acrylamide cross-linker while maintaining a constant total acrylamide concentration. Macrophages showed a six-fold preference for stiff over soft opsonized particles. Observations suggest that phagocytosis of soft particles is inhibited downstream of receptor binding. Staining of actin with rhodamine phalloidin showed that actin filaments were polymerizing beneath 85% of the stiff particles, but only under 11% of the soft ones,

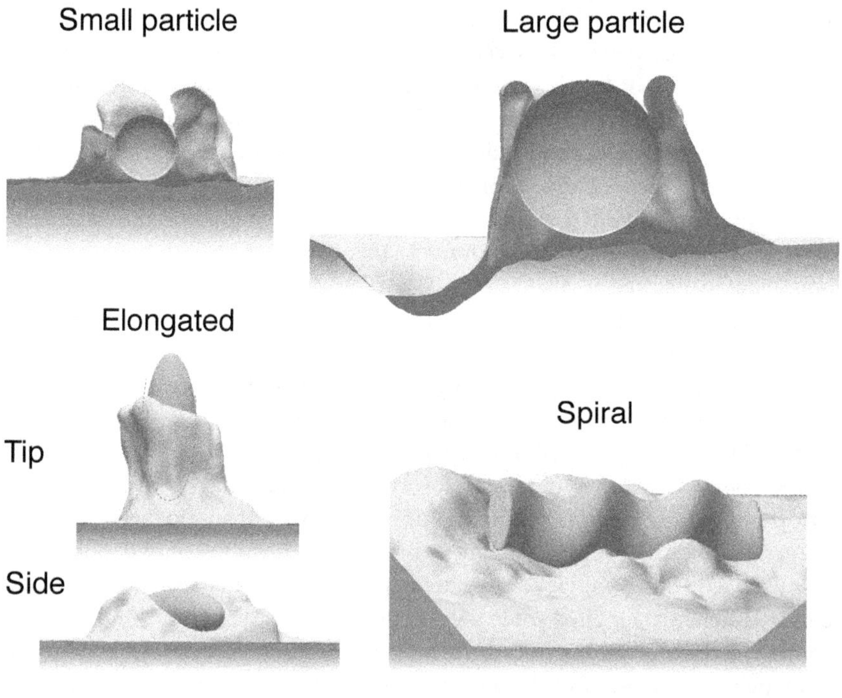

Figure 6.3 Engulfment is constrained by particle geometry. Simulations show phagocytosis is weakly dependent on particle size with both small (radius 1.2 μm, top left panel) and large (radius 3 μm, top right panel) particles. However, simulations show uptake is strongly dependent on particle shape and orientation. An elongated particle is easily taken up tip first, but not by the long axis (bottom left panel). Spiral-shaped particles are not taken up (bottom right panel). Simulation time is comparable in all panels. Image adapted from Tollis *et al.* (2010).

despite the formation of adhesive sites beneath all particles regardless of stiffness. This lack of actin organization is probably responsible for the failure of cells to engulf soft particles. The authors proposed that impaired actin polymerization might be due to disruption of the Rho GTPase signalling pathway. To test the involvement of RhoA and Rac1 in mechanosensing they microinjected constitutively active L63RhoA and L61Rac1 into macrophages prior to the presentation of hard or soft particles. Constitutively active Rac1 enhanced ruffling and lamellipodial progression, and restored the ingestion of soft particles. In agreement with this, microinjection of dominant negative N17Rac1 inhibited the uptake of hard particles. Phagocytosis of soft particles was also restored by treating particles with lysophosphatidic acid (LPA), an activator of several signalling pathways including the Rho GTPases. LPA stimulated actin polymerization beneath soft particles and restored engulfment of these particles without enhancing phagocytosis of hard particles. Therefore, rather than activating an independent, indiscriminate mechanism of phagocytosis, LPA probably elicits a signal normally stimulated by the rigidity of the opsonized target.

The strong influence that particle stiffness has on phagocytic engulfment is similar to observations on cell migration. Specifically, the study of fibroblasts migration on 2D substrates revealed that cells exhibit a preference for a stiff substrate (Lo *et al.*, 2000). In accordance with these results, carcinoma cells were shown to move along mechanical stiffness gradients towards the region of maximal stiffness. This shows that cells prefer to move on stiff environments (Hong *et al.*, 2007).

Role of membrane availability

The most striking observation about phagocytic cells is that the plasma membrane is highly deformed to allow engulfment of the target particle. If the particle is large, significant amounts of membrane need to be recruited or produced, at least locally (Cannon and Swanson, 1992; Holevinski and Nelson, 1998). Indeed, stretching can be ruled out on the basis of the physical properties of lipid bilayers, which can only be stretched by a few per cent (Bivas *et al.*, 1997). The amount of material that can be internalized by a single phagocyte is remarkable: human neutrophils, with a diameter of almost 10 μm, can internalize spherical particles as large as 8 μm (Champion and Mitragotri, 2006; Herant *et al.*, 2006). In these assays, the neutrophil membrane extends locally around the particle but the cell body seems to retain its spherical shape and original dimensions. Therefore it is obvious that the surface area of the cell must expand for phagocytosis to occur.

The large extension of membrane during phagocytosis was demonstrated by measuring the membrane capacitance of human monocyte-derived macrophages and cells from the murine macrophage-like cell line J774.1. Membrane capacitance is directly proportional to the cell membrane surface area, and therefore provides an indirect way to monitor the increase in membrane area during phagocytosis. When cells are exposed to particles ranging from 0.8 to 3.2 μm, the membrane capacitance decreased by an average of 8 pF per cell. Interestingly, the capacitance decrease was found to be quantized in units of 250 fF for particles 0.8 μm in diameter, and 480 fF for particles 3.2 μm in diameter, with each unit corresponding to the engulfment of an individual particle (Holevinski and Nelson, 1998). As expected, engulfing larger particles requires more membrane, which translates into a larger decrease in membrane capacitance. In other experiments, dynamin 2, a GTPase involved in the scission of newly formed vesicles from membrane compartments and their transport and fusion with another compartment, has been shown to control vesicle exocytosis occurring during phagocytosis in macrophages, although the intracellular source of membrane was not identified (Di *et al.*, 2003). In *Dictyostelium*, blocking intracellular membrane traffic leads to a strong decrease in cell motility, due to impaired cell protrusion (Thompson and Bretscher, 2002; Zanchi *et al.*, 2010).

Several sources of extra membrane have been experimentally observed. First, some cell types, such as neutrophils, have numerous wrinkles and folds, which provide a natural membrane reservoir (Herant *et al.*, 2005; Hallett and Dewitt, 2007). However, complete unfolding of slack wrinkles requires an additional force, as the wrinkles are held together by a molecular 'velcro-like' mechanism (Herant *et al.*, 2003). During phagocytosis the force required to detach the wrinkles is significantly reduced (Herant *et al.*, 2005). The membrane from the wrinkles would then become available for phagosome formation. Furthermore, it was recently shown that during phagocytosis macrophages loose their ability to generate random actin-rich membrane protrusions far away from the cup. This observation was made when studying how phagocytosis of the fungi *Cryptococcus neoformans* affected the macrophage response to chemoattractants such as fractalkin (FKN) and colony-stimulating factor 1 (CSF-1) (Luo *et al.*, 2009). The chemotactic response was estimated by measurement of the formation of actin-rich membrane protrusions (ruffles), produced in response to stimulation with FKN or CSF-1. The response was increasingly inhibited the more pathogens were internalized by phagocytosis. This observation was partially attributed to the lack of membrane resources.

Second, internal sources provide the nascent phagosome with new membrane (Goodall *et al.*, 1972). Grinstein and colleagues showed that the SNARE (soluble *N*-ethylmaleimide-sensitive factor attachment protein receptor) complex was necessary for phagocytosis to proceed, and the function of SNARE is to facilitate membrane fusion. This supports the notion that internal membranes are recruited to and fuse with the plasma membrane at the site of phagocytosis, a process called 'focal exocytosis' (Booth *et al.*, 2001). Several intracellular compartments contribute to the formation of the phagosome, including early and late endosomes (Bajno *et al.*, 2000; Lee *et al.*, 2007), as well as the endoplasmic reticulum (ER; Gagnon *et al.*, 2002; Ackerman *et al.*, 2006). For instance, Rab4 and Rab11 on recycling endosomes are also present on phagosomes. In *Dictyostelium*, the ER proteins calnexin and calreticulin are localized at sites of particle engulfment, and a double knock-out is strongly impaired in phagocytosis (Müller-Taubenberger *et al.*, 2001). Moreover, a large body of proteomic data reveals the presence of ER proteins in the phagosome proteome (Desjardins, 2003; Boulais *et al.*, 2010; Dieckmann *et al.*, submitted). In some cases, lysosomes have been shown to fuse with the plasma membrane during phagocytosis. In addition to providing membranes, these internal compartments also deliver specific proteins that build the phagosome and initiate its maturation (Desjardins *et al.*, 1994 ; Peyron *et al.*, 2001).

Focal exocytosis of intracellular membranes relies on several steps, including vesicle budding from a donor compartment, transport along cytoskeletal tracks and fusion of vesicle with a receptor compartment (Niedergang and Chavrier, 2004). Dynamin appears to be involved in the production and/or movement of intracellular vesicles to the cell surface to support membrane expansion during phagosome formation (Di *et al.*, 2003). Similar to the situation in mammalian cells, in *Dictyostelium* Dynamin is found on early phagosomes (Dieckmann *et al.*, 2010) and its knock-out leads to impaired phagocytosis (Wienke *et al.*, 1999). Evidence for membrane delivery at the phagocytic cup is also suggested by the fact that knocking out a NDR family kinase that interacts with the vesicular associated protein

EmpC slows down uptake of large particles (Kastner *et al.*, 2010). Another study points out that the ER itself can fuse with the plasma membrane during phagocytosis (Gagnon *et al.*, 2002; Becker *et al.*, 2004).

The need for sources of membrane is reflected in the quantitative results of the stochastic model introduced above (Fig. 6.4). Membrane parameters corresponding to realistic cup shapes were estimated. As a result, the membrane surface tension was found to be more than one order of magnitude lower than the tension reported in the literature from bulk measurements. Although it is argued that membrane tension may be locally down-regulated at the cup, it is likely that phagocytic engulfment of particles with diameter larger than 1 μm is impossible without delivery of new membrane at the growing cup.

Role of actin in force generation

Although phagocytosis generally involves active processes such as actin polymerization in the cup (May and Machesky, 2001), recent reports indicate that phagocytosis may still work in an actin-independent manner. Indeed, phagocytic uptake was observed despite treating phagocytes with cytochalasin D (Van Zon *et al.*, 2009). Hence, ligand–receptor binding

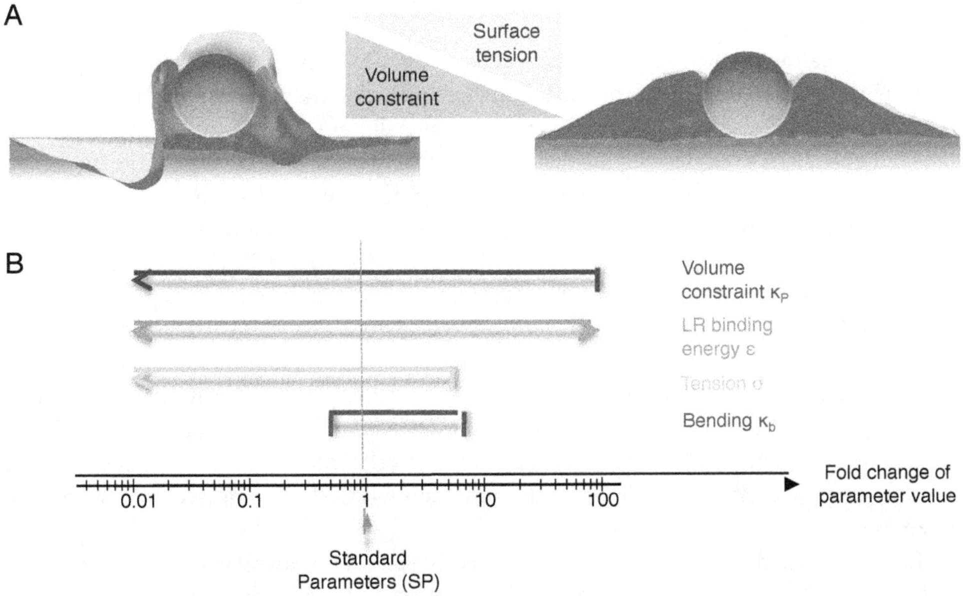

Figure 6.4 Dependence of cup shape on cell-membrane parameters. (A) Influence of surface tension and volume constraint. Simulated phagocytic cup for (left) a large volume constraint ($\kappa_P = 10^{-2}$ pN/μm^5) and a relatively low surface tension ($\sigma = 6.2 \times 10^{-7}$ mN/m), and (right) a high surface tension ($\sigma = 6.2 \times 10^{-5}$ mN/m) and a nearly unconstrained cell volume ($\kappa_P = 3.8 \times 10^{-7}$ pN/μm^5). Both sets of parameters produce approximately the same speed for engulfment. (B) Ranges of parameter values for successful engulfment. Simulation time was restricted to twice the engulfment time with Standard Parameters (SP) (Tollis *et al.*, 2010) and represented by vertical dashed line. Shown are fold changes of bending stiffness (black pointed arrow), surface tension (grey double pointed arrow), ligand–receptor energy density (light grey pointed arrow) and volume constraint (black blunt arrow) along x-axis. Pointed arrows indicate that parameter regimes may extend beyond tested limits, blunt arrows indicate sharp limits beyond which full engulfment is not reached in simulation time. Image adapted from Tollis *et al.* (2010).

alone may be sufficient to guide the cell membrane around the particle. From now on, we will call this mechanism the passive zipper, in contrast to the active zipper where actin acts to stabilize ligand–receptor bonds. This passive uptake hypothesis was tested with a combination of simulations and experiments (Tollis *et al.*, 2010).

To investigate the energetic requirements of the zipper mechanism, the authors implemented simulations of the passive zipper, in which ligand–receptor bonds are not stabilized by actin polymerization and hence remain reversible. They argue that engulfment may still progress if the energetic cost of stretching and deforming the membrane is offset by the ligand–receptor binding energy in the presence of thermal membrane fluctuations. As a result, engulfment of small particles (1.5 μm radius) by the passive zipper is possible, albeit more slowly and via more variable phagocytic cup shapes than engulfment by the active zipper (Fig. 6.5). In contrast, large particles (3 μm radius) are not engulfed by the passive zipper. For the active zipper, random membrane fluctuations are rectified by irreversible ligand–receptor binding due to actin polymerization. This leads to uniform progression of the cell membrane all around the particle at approximately the same speed. In contrast, engulfment by the passive zipper occurs through binding of large membrane ruffles, which eventually enclose the particle. The variability of the phagocytic cup could therefore be envisaged as a measure of the respective contributions of active and passive processes in engulfment progression.

This simulation prediction was successfully tested experimentally. To specifically compare active with passive engulfment for small and large particles, the authors transfected COS-7 cells with either wild-type FcγR (WT-FcγR) or a signalling-dead mutant receptor (Y282F/Y298F-FcγR). Cells expressing the wild-type receptor are expected to perform active engulfment, whereas cells expressing the signalling-dead mutant receptor are expected to perform passive engulfment. As a control, passive engulfment is additionally implemented by treating cells expressing WT-FcγR with 0.2 μM cytochalasin D (WT-FcγR+CytoD), which prevents actin polymerization. Synchronized phagocytosis assays using small (1.5 μm radius) and large (3 μm radius) IgG-opsonized polystyrene particles were carried out, and, after fixation, receptor localization in phagocytic cups was visualized by fluorescence confocal microscopy. Reconstruction of 3D phagocytic cups from confocal microscopy data revealed a trend of the variability in cup shape consistent with our model: at least for partially engulfed small particles the variability increased for cells transfected with signalling-dead mutant receptor or WT-Fcγ+CytoD cells compared with cells transfected with wild-type receptor.

This study hence demonstrated that passive uptake, i.e. not involving actin polymerization, is sufficient to ingest small particles. During the subsequent evolution of this system there may have been a selective advantage in being able to phagocytose larger, more complex shaped particles in a more robust manner. This may have led to phagocytosis as we know it today with complex regulatory pathways and active uptake processes.

Although in good general agreement with the experimental result, the minimal model of the zipper mechanism discussed above shows some discrepancies with observations, especially during the later stages of engulfment. Specifically, computer simulations show slow and disordered cup closure, whereas experimentally imaged cups close quickly and regularly through thin lamellipod-like extensions. The latter features of the closing cups indicates that contractile forces generated at the leading edge of the cup by myosins in the actin-rich ring are important for shaping and closing the cup.

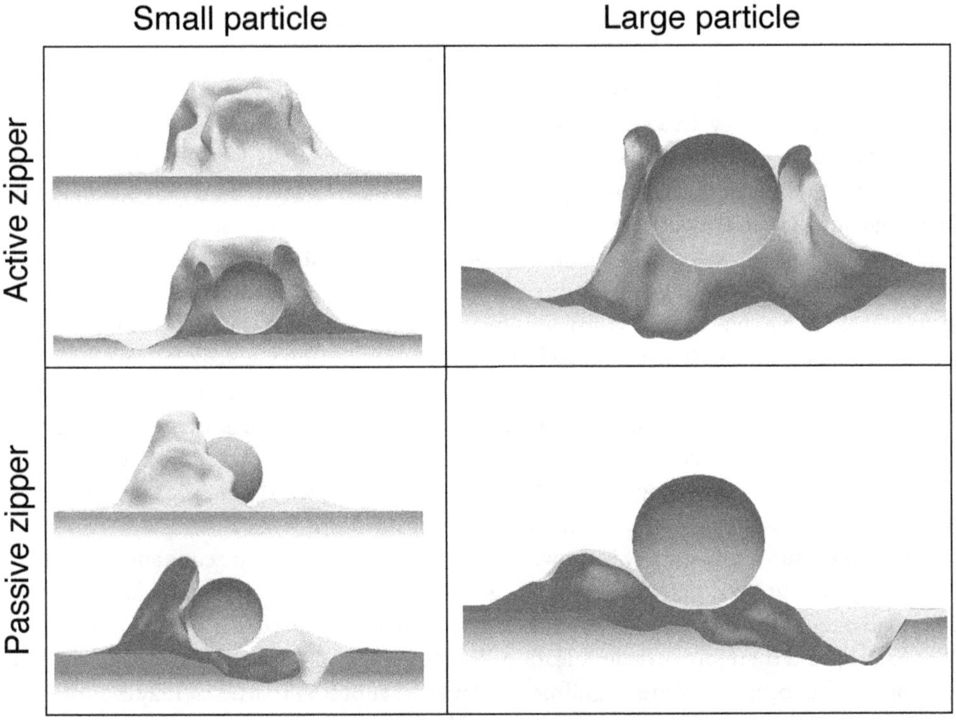

Figure 6.5 Simulations showing engulfment by active and passive zipper mechanisms. Top left panel: Side view (top) and cross section (bottom) of a phagocytic cup obtained for the active zipper and a small (3 μm of diameter) particle. Top right panel: Cross-section of a cup obtained for the active zipper and a large (6 μm of diameter) particle. Bottom left panel: Side view (top) and cross-section (bottom) of the cup obtained for the passive zipper and a small particle (same overall simulation time than top left panel). Bottom right panel: Cross-section of a cup obtained for the passive zipper and a large particle (same simulation time than top right panel). Image adapted from Tollis *et al.* (2010).

Role of myosins in contraction and cup closure

Myosins are actin-associated motor proteins with known or predicted roles in many types of eukaryotic cell behaviour, including endocytosis and phagocytosis (Soldati and Schliwa, 2006), cell adhesion, migration, and division (cytokinesis). These proteins are able to couple hydrolysis of ATP to conformational changes that result in their movement on actin filaments. The overall structure of myosins consists of a well-conserved N-terminal motor (head) domain, a neck domain and a more variable C-terminal tail domain. The latter is used for regulating signalling and for binding a variety of effector proteins or lipids. Single-headed class I myosins were shown to localize to regions of dynamic actin turnover and play a role in several endocytic processes, including phagocytosis (Soldati and Schliwa, 2006). Double-headed class II myosins contribute to maintenance of the actin cortex and effectively regulate the surface tension. For instance, if a two-headed myosin binds two anti-parallel actin filaments, the myosin motion generates a movement of actin filaments relative to each other. In an actin bundle or an actin ring, sliding of actin filaments with respect to each other leads to the contraction of the bundle or ring.

The role of myosins in cup closure was demonstrated by Araki and collaborators (Araki *et al.*, 2003). Specifically, the authors performed phagocytosis assays with cells treated with ml-7, a myosin light-chain kinase inhibitor that perturbs phagocytic cup closure but not actin filament assembly. In a ml-7-treated cell, F-actin assembly and phagocytic cup extension occurred, but the process was slower and the cups were shallower than in the control cells without inhibitor. Even though phagocytic cups were formed, they were inefficient at keeping contact with the particle and closing.

Several years ago, Swanson and collaborators demonstrated that myosin IC was involved in constriction forces during cup closure (Swanson *et al.*, 1999). Time-lapse microscopy of Fc-mediated phagocytosis of erythrocytes by mouse macrophages revealed strong contractile activities when one IgG-opsonized erythrocyte was simultaneously taken up by two macrophages. At the junction between the two advancing cups, the erythrocyte was pulled into two phagosomes, only connected by a string of erythrocyte membrane. Among various myosins present in the phagocytic cups, the authors reported that only myosin IC was concentrated around the strings connecting shared erythrocytes, suggesting that myosin IC mediates the purse-string-like constriction that closes cups.

In another study, myosin X was shown to be recruited to phagocytic cups and expression of a truncated myosin X tail inhibits the phagocytosis of large particles (6 μm diameter) but not small ones (2 μm diameter) (Cox *et al.*, 2002). Myosin X binds to PI(3,4,5)P_3 through its PH domain, and therefore myosin X leads to F-actin/membrane contractions. The fact that inhibition of myosin X prevents engulfment of large, but not small particles, may indicate two possibilities: either the particle size is sensed by the cell to recruit only minimal resources for particle engulfment, and therefore actin–myosin contractions are not used for such small particles, or small particles are completely engulfed before actin–myosin contractions begin.

Bistability, stochasticity and decision-making

Recent experimental results have raised fundamental questions about cell decision-making in the context of phagocytosis and its implications for the zipper mechanism. The apparent absence of cellular decision-making during the onset of phagocytosis (which would support the trigger mechanism) is currently a matter of great debate, as engulfment was found to either stall or complete later in the process (Van Zon *et al.*, 2009; Tollis *et al.*, 2010; Zhang *et al.*, 2010). In the original work on the zipper mechanism, particle-to-particle variability was already apparent and mentioned by the authors (Griffin *et al.*, 1975, 1976). Indeed about 10 particles were attached to the macrophages but only two or three on average were completely engulfed. This difference between attachment and phagocytosis demonstrates that stochasticity is present at the cellular level. Why are some particles completely ingested whereas some others remain only attached? Does the cell somehow choose what particles to ingest, or is there a competition for resources, leading, as a consequence of molecular stochasticity, to a large variability in particle fate? The former would correspond to cellular decision-making, and therefore to triggering of phagocytosis in contradiction to the zipper mechanism. The latter situation would be possible, if the zipper mechanism leads to bistability. Slight local fluctuations in molecular patterns would either stall engulfment, or lead to a continuously progressing cup. Recently, several experimental and modelling studies were conducted to shed light on these questions.

To quantitatively compare the zipper and trigger mechanisms, Swanson and collaborators have developed an experimental technique that measures specific intracellular responses at various stages of engulfment, as a function of IgG density on the particle (Zhang *et al.*, 2010). The authors tagged specific molecules involved at different stages of the phagocytic pathway with yellow fluorescent protein (YFP) and challenged macrophages with polystyrene particles, coated with different densities of IgG. Then they calculated the fluorescent signal corresponding to the molecules specifically recruited to the phagocytic cups, measuring therefore the exact cellular response to the FcR/IgG binding stimuli (assumed to be proportional to IgG density) with the number of FcR in excess. The authors managed to distinguish three stages of uptake: (1) early engulfment, during which signalling is dominated by Syk (tagged to measure the initial response), (2) intermediate engulfment level, during which actin is widely polymerized throughout the cup (actin, and p85, a regulatory domain of PI3K, were chosen to label the cellular response during this stage), and (3) late engulfment, during which Abelson Tyrosine Kinase (AktPH; colocalizes with PI(3,4,5)P_3) and PKC were tagged.

During early engulfment, particles with ten-fold higher IgG density stimulated a 10-fold larger Syk response in the cells, consistent with the zipper mechanism and independent receptor signalling. Indeed, if each receptor bound to a ligand generates its own Syk response, the total cellular response will be proportional to IgG density. At the intermediate stage, molecules downstream of Syk (e.g. PI3K and F-actin) begin to show a non-linear response to the IgG signal: particles with 10-fold higher IgG density recruit only 30% more of these secondary molecules, showing either negative receptor cooperativity, or, most likely, non-linear feedback in the pathways downstream of Syk. Non-linearities are, in principle, not in contradiction with the zipper mechanism. Indeed, if downstream feedback mechanisms regulate the recruitment of active molecules, and therefore quantitatively modify the amount of each molecular species at the cup, it is possible to reconcile this result with the zipper mechanism.

Later stages of engulfment show even more non-linear signatures in the response. AktPH recruitment, and therefore PI(3,4,5)P_3 localization at the cup, is surprisingly similar for low and high IgG density particles. Furthermore, independent of IgG density, AktPH is significantly less in stalled cups than in successfully proceeding cups, indicating that PI3K amplification and PI(4,5)P_2 to PI(3,4,5)P_3 conversion correlates with successful advancement of the phagocytic cup. Even more strikingly, PKC shows an inverse relationship to IgG density, i.e. low density particles attract more PKC at cups. The absence of PKC correlates with stalling of cup progression, whereas PKC is largely amplified in successful cups. According to the authors, this indicates that by mobilizing the PI(3,4,5)P_3 machinery, the cell actually makes the decision to complete engulfment: the zipper mechanism drives early engulfment, and at some critical stage in cup progression, late signals trigger engulfment completion. By inhibiting PI3K (with LY294002), the authors examined the role of PI(3,4,5)P_3 in the cell's commitment to phagocytosis by modulating the cell's ability to generate PI(3,4,5)P_3. Actin polymerization is similar in both stalled and advancing cups at the same level of engulfment, indicating that actin polymerization is independent of PI(3,4,5)P_3. However, downstream protein PKC is undetectable in stalled cups, indicating that PI(3,4,5)P_3 correlates with successful engulfment. Therefore, a threshold in PI(3,4,5)P_3 concentration has to be reached to allow further engulfment, and exceeding this threshold requires a sufficient amount of FcR/IgG bonds. Hence, in contrast to the zipper mechanism, engulfment completion is triggered by a suprathreshold amount of PI(3,4,5)P_3, accumulated in the cup by the previous sequential ligand–receptor binding. Triggering, or cell-decision, arises at least 4 minutes before the end of engulfment.

There is however another interpretation of these results. Syk recruitment, throughout engulfment, depends linearly on IgG density. As a consequence, in all engulfment stages, primary signals immediately downstream of FcR are linearly correlated with IgG density, in complete accordance with the zipper mechanism. Primary signals are altered, regulated, and fed back by the large number of interconnected signalling pathways, each of them having different characteristic time constants. During early engulfment, only the regulatory pathways with short characteristic times are relevant. However as engulfment progresses, more and more regulatory mechanisms become involved, adding up more and more sources of non-linearity in the response, until the most non-linear response (the all-or-none response) is reached during late engulfment. In other words, the initial zippering may or may not be able to accumulate enough $PI(3,4,5)P_3$ in order to allow signal transition and engulfment completion after the last checkpoint.

Bistability, an all-or-none outcome for phagocytosis, was also observed by Van Zon and co-workers (Van Zon *et al.*, 2009). By combining mathematical modelling with phagocytosis assays using transfected COS-7 cells and opsonized polystyrene particles, the authors studied the distribution of phagocytic cup sizes at different times. At late times the distribution was reported to be bimodal with a broad distribution of cup sizes around half engulfment and a sharp peak at full engulfment that increased with time. Cups with intermediate engulfment were rarely observed. This gap in the distribution implies that phagocytic cup growth is highly stochastic, and that cell-to-cell or even particle-to-particle variability is high. At that time, this result was interpreted in terms of a mechanical bottleneck at half-engulfment: in their mathematical model, the rate of growth of the phagocytic cup size is mainly determined by the balance between actin-generated forces, which tend to push out the cell cortex and membrane for engulfment, and the resistance of the cortex and membrane to deformation. In their model, the force of the membrane/cortex to resist its deformation increases during the first half of engulfment before reaching a plateau at half engulfment. Intuitively, we expect that there is a maximal pro-engulfment force that can be locally produced by polymerizing actin. If this maximum actin-generated force is smaller than the resistance force, phagocytosis will stall before it reaches half-engulfment. However, if it is larger, the phagocytic cup will fully complete. Now, Swanson's work provides an alternative explanation for the bimodal distribution of cup sizes in terms of non-linear signalling.

The mechanical bottleneck idea, although convincing, originated from a deterministic model, making a discussion of bistability or decision-making difficult. Deterministic models have been very useful for describing force generation by the cytoskeleton, however they are unable to fully address why uptake depends on particle-shape, as they assume rotational symmetry around the axis connecting cell and particle (Herant *et al.*, 2005; Herant *et al.*, 2006; Van Zon *et al.*, 2009). Additionally, large particle-to-particle variation in cup growth (Van Zon *et al.*, 2009) and cell-to-cell variation in the related process of endocytosis (Snijder *et al.*, 2009) point towards the importance of stochasticity during the uptake, which was not captured in previous deterministic approaches.

In addition to reproducing the dependence on particle-shape the stochastic model of the zipper mechanism also reproduces this bimodal distribution of phagocytic cup sizes (Fig. 6.6; Tollis *et al.*, 2010). Slight (much less than one order of magnitude) variation of the surface tension model parameter, produced either cups stalled at the half-engulfed stage, or completely finished cups. Owing to the stochastic nature of the model, the two different outcomes were also occasionally obtained for the same parameter values. The high sensitivity to parameter changes explains the particle-to-particle variation observed in experiments. However, this model is also not fully suited to discuss bistability and decision-making, owing to the absence of key signalling components in the model, and further research will be required.

Live-cell imaging of phagocytosis in amoeba

Space-time coordination of signals

Signalling during phagocytosis appears to be highly coordinated in space (along cup) and time, ranging from seconds for signalling to minutes for full engulfment (Swanson and Hoppe, 2004; Swanson, 2008). Furthermore, signals from individual receptors need to be integrated to recognize different particle shapes and to decide whether to engulf or not. Specifically, signals produced by ligand–receptor complexes pass through different

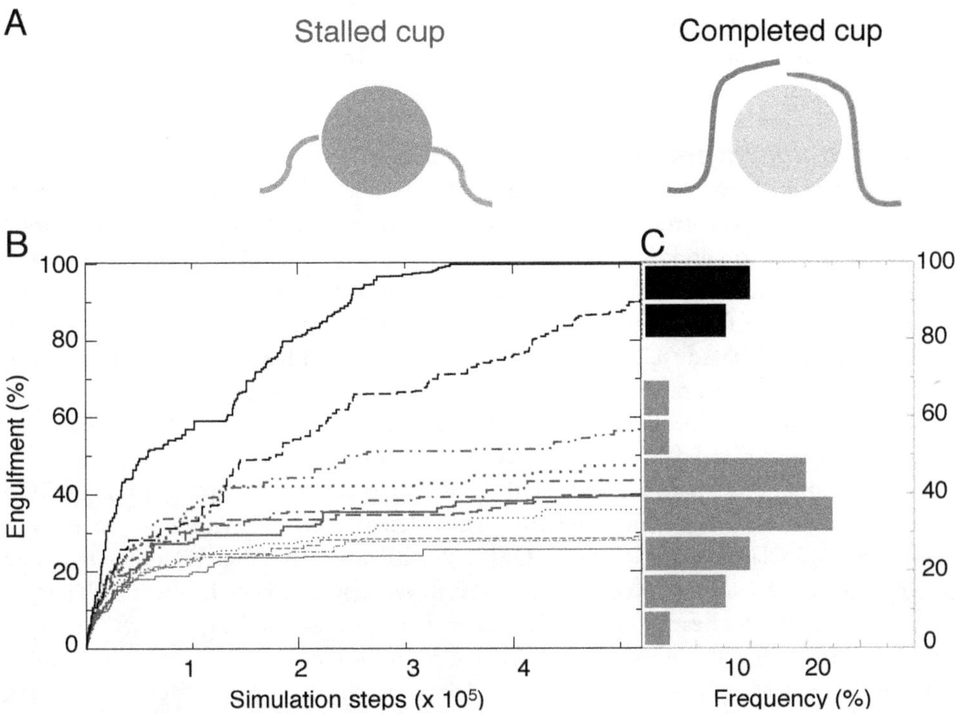

Figure 6.6 Stalling at partial engulfment. (A) Schematics of a stalled cup (left) and a completed cup (right). (B) Time course of the percentage of engulfment as a function of increasing surface tension in simulations. Shown are completed (or nearly completed) uptake (top solid and dashed lines) and partial (stalled or very slowly progressing) uptake (other lines). Surface tension values ($\sigma \times 10^{-4}$ mN/m) used are 0.8 (top solid), 1.1 (top dashed), 1.4 (thick dash-double dotted), 1.6 (thick dotted), 1.9 (thick dashed-dotted), 2.1 (thick dashed), 2.3 (thick solid), 2.8 (thin dotted), 3.1 (thin dash-dotted), 3.4 (thin dashed), and 3.7 (thin solid). (C) Distribution of engulfed particle surface area measured in experiments with COS-7 cells transfected with wild-type receptors (WT-FcγR) after 10 minutes. Black and grey bars correspond to the two peaks of a bimodal distribution. Image adapted from Tollis *et al.* (2010).

stages, involving diffusible molecules. Particular spatial patterns of, for example GTPases and phosphoinositides, may be key in determining successful engulfment. Understanding the mechanics of phagocytic engulfment requires a complete understanding of the signals, which lead to membrane deformations. Therefore, real-time space-resolved data obtained from live cells and modelling of such are indispensable for the next generation of results on phagocytosis.

In recent years, drastic improvement of imaging techniques together with a wide range of fluorescent probes allowed researchers to address the dynamics of actin remodelling and signalling in live cells. In this part of the chapter, we will mainly focus on the contribution of live-cell imaging to the dissection of phagocytosis in the model organism *Dictyostelium*. Indeed, in recent years studies on this professional phagocyte have considerably improved our understanding of basic mechanisms of phagocytosis. Here, we will review two aspects of phagosome formation. We will first focus on the genesis and spatial distribution of membrane territories, especially of phospholipids during cup formation, before addressing the spatiotemporal aspects of myosins and the actin cytoskeleton. Subsequently, we will also discuss the modulation of these processes during the engulfment of small and large particles, as well as particles with complex shapes. These specific aspects were pioneered in the study of *Dictyostelium* and may have predictive value for the understanding of phagocytosis in immune cells.

Spatio-temporal dynamics of phospholipids

Dormann and coworkers used live imaging of *Dictyostelium* cells expressing a panel of PH domain – GFP constructs to characterize the dynamics of phosphoinositides during phagosome formation and especially PI(3,4,5)P_3 and PI(3,4)P_2 (Dormann *et al.*, 2004). Upon particle binding and initiation of cup formation, PI(3,4,5)P_3 accumulates and stays associated with the forming phagosome until completion of internalization. Formation of PI(3,4,5)P_3 correlated well with actin assembly. Indeed, PI(3,4,5)P_3 is able to recruit SCAR/WAVE, an activator of the Arp2/3 complex. As soon as the phagosome is formed, PI(3,4,5)P_3 is rapidly hydrolysed into PI(3,4)P_2. This timing correlates well with dissociation of the actin coat and delivery of the v-ATPase (Clarke *et al.*, 2010b).

An interesting observation is that the time of association with the phagosome of GFP-CRAC, a probe for PI(3,4,5)P_3 and PI(3,4)P_2, depends on the nature of the particle (Giorgone and Clarke, 2008). Indeed, GFP-CRAC dissociated from bacteria-containing phagosomes shortly after internalization, while it stayed associated much longer with phagosomes containing latex beads or silica microspheres, suggesting that the nature of the particle influences early signalling events via differential receptor binding. The authors also noted that delivery of the v-ATPase was somewhat delayed in the case of phagosome containing latex beads compared with bacteria, although this observation was not strictly quantified. Past the rapid acidification phase, the main phosphoinositide present on the phagosomal membrane is PI(3)P (Fig. 6.7). The *Dictyostelium* arrestin-related protein AdcA has been shown to localize to the phagosome via its PI(3)P binding FYVE domain were it might modulate signalling of G-protein-coupled receptors (Guetta *et al.*, 2010).

Actin-wave phenomenon

Work in *Dictyostelium* illustrated the very intricate and dynamic relation between the actin cytoskeleton and phosphoinositides. Indeed, in an elegant study, Gerisch and coworkers

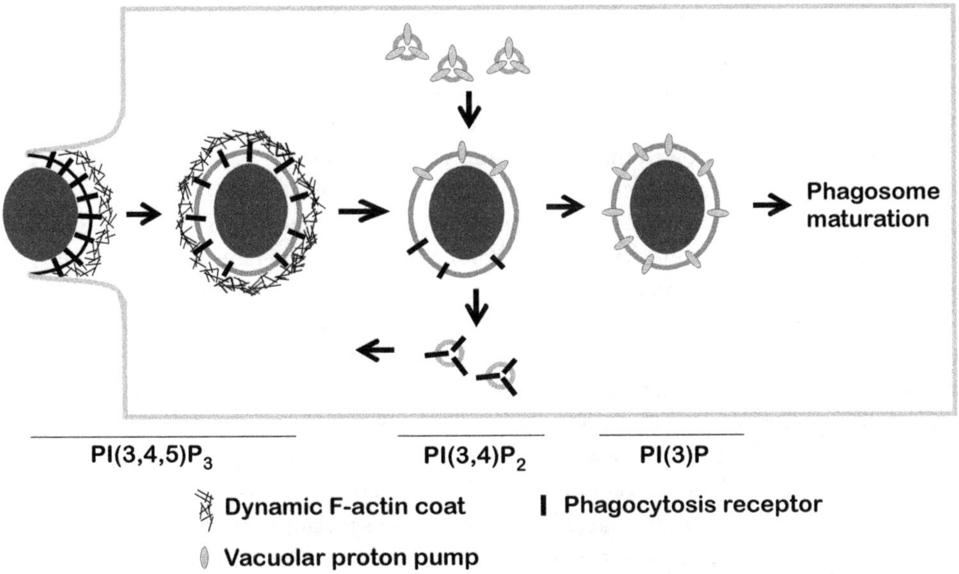

Figure 6.7 Phagocytosis in *Dictyostelium*. After recognition and binding, the particle is internalized. Rapidly after internalization, the actin coat is disassembled and the phagosome is acidified by delivery of the vacuolar proton pump. Concomitantly, plasma membrane proteins are recycled away from the phagosome. These processes are accompanied by distinct changes in phosphoinositide composition.

showed that actin waves travel along the planar membrane of substrate-attached cells (Gerisch *et al.* 2009). Total internal reflection fluorescence (TIRF) as well as confocal microscopy of cells expressing an mRFP–LimEΔ fusion protein, which labels newly synthesized filamentous actin, visualized highly motile actin waves travelling along the ventral plasma membrane. Use of a PHcrac-GFP construct highlighted that these waves co-localized with the edge of a PHcrac-GFP positive area, independently of its direction of propagation and of its expansion or retraction phase. Moreover, a GFP-TAPP construct that specifically recognizes $PI(3,4)P_2$ showed no difference in labelling of the inner or external area of the wave indicating that $PI(3,4,5)P_3$ is the species that distinguishes both area.

The study also revealed the interdependent nature of $PI(3,4,5)P_3$ and actin polymerization. Indeed, treatment of the cells with latrunculin A, a drug that prevents actin polymerization, prevented both wave propagation and $PI(3,4,5)P_3$ formation. Similarly, $PI(3,4,5)P_3$ formation and actin waves were inhibited by treatment with the PI3K inhibitor LY-294002. This organization and dynamics of $PI(3,4,5)P_3$ and actin waves are highly reminiscent of their relationships in a phagocytic cup, as observed by live imaging of *Dictyostelium* cells engulfing long, rode-shaped particles. In addition, resting cells harbouring high actin wave activity had a higher phagocytosis rate. The authors suggest that dynamic actin waves might be a way to probe the environment and prepare the cell to phagocytose. Once a particle is encountered, receptors are engaged and the dynamic wave is then reorganized into a stable structure.

To ensure proper digestion of the particle by lysosomal enzymes, acidification of the phagosomal lumen is probably the most conserved process. The delivery of the vacuolar proton pump (v-ATPase) has been well studied in *Dictyostelium* (Clarke *et al.*, 2002, 2010b). Studies by live microscopy using a VatM-GFP construct have shown that the v-ATPase is delivered during or right after particle uptake. The newly formed phagosome rapidly fuses with endosome harbouring the v-ATPase in their membrane. Pre-labelling of endosomes with TRITC-Dextran showed that most of the v-ATPase positive vesicles were part of the endolysosomal system and possibly only a minor part is delivered from the biosynthetic pathway. Altogether, data suggest that membrane delivery at the site of phagosome formation is a highly conserved mechanism. Concomitantly with membrane delivery, membrane protein sorting and exclusion mechanisms are also crucial to determine membrane composition and identity during phagosome formation and maturation in *Dictyostelium* (Mercanti *et al.*, 2006) and in mammalian phagocytes (Lee *et al.*, 2007). Such processes are probably co-ordinated with the formation of functional membrane territories often delineated by various phosphoinositide isoforms (Gerisch *et al.*, 2009).

As mentioned earlier, macrophages continuously probe their environment by extending thin actin rich protrusions that require Rac and PI(3,4,5)P_3 as well as PI(4,5)P_2 (Flannagan *et al.*, 2010). The authors argue that efficient probing is especially important for the recognition and binding of scarcely opsonized particles. Both the circular waves and the finger-like probing mechanism seem to depend on the same molecules. While actin waves might be a general way to scan the environment for non-specific uptake of particle, a more fine-tuned mechanism might be necessary for macrophages to search for more specific targets.

Differential myosin localization

Dictyostelium expresses seven class I myosins, three long tails or amoeboid myosins (MyoB/C/D), three short tail (MyoA/E/F) and one divergent (MyoK). A conserved feature of class I myosins seems to be the overall organization of their binding partner networks. Indeed, several class I myosins are able to recruit, directly or indirectly, the Arp2/3 complex and proteins associated with dynamic actin turnover. This is the case for example for *Dictyostelium* MyoB and MyoC that bind CARMIL, an adaptor protein that binds and activates the Arp2/3 complex (Jung *et al.*, 2001) as well as for MyoK that recruits the Arp2/3 complex via interaction with Abp1 (see Soldati, 2003, for review). Class I myosins are also able to bind membranes via their tail domain, a key function in membrane deformation processes such as phagocytosis and macropinocytosis, as illustrated by myosin IC's role in the purse-string-like contraction generated during cup closure, as mentioned earlier in this chapter (Swanson *et al.*, 1999).

A more recent study using live microscopy and immunolabelling of fixed cells (Dieckmann *et al.*, 2010) dissects in a highly detailed fashion the localization and role of class I and II myosins as well as their binding partners during uptake of large particles. Feeding budded yeast to *Dictyostelium* cells results in the formation of a phagocytic furrow at the level of the yeast neck (Figs. 6.8 and 6.9). Engulfment of the hourglass-shaped particle results in the transient stalling of the phagocytic zipper machinery at the bud neck induced by the inversion of curvature as illustrated the localization of MyoB-GFP during uptake of a budded yeast (Fig. 6.8).

Taking advantage of this phenomenon, the authors were able to localize components of the actin polymerization machinery with unprecedented spatial resolution. While myosin II is mainly localized at the plasma membrane, class I myosins appeared to be mainly associated with the forming phagosomal membrane. Analysis of the phagosomal machinery at the furrow allowed the observation of subterritories of actin binding proteins. The proximal phagosomal membrane-associated class I myosins were followed by concentric and overlapping layers of Abp1, Arp3 and coronin, and finally the more distal layer occupied by myosin II. This order is strikingly reminiscent of the overall organization of the lamellipodium of migrating cells, suggesting that force generation at the plasma membrane is a highly conserved mechanism.

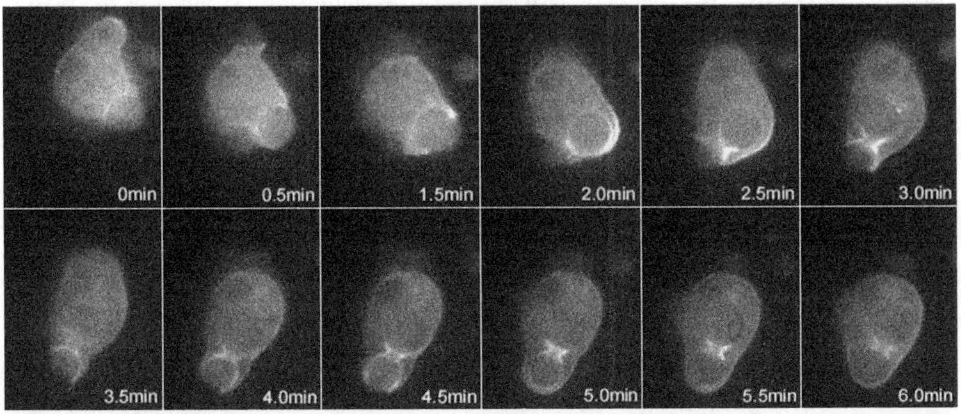

Figure 6.8 Live imaging of uptake of a budding yeast by *Dictyostelium* cells expressing MyoB-GFP.

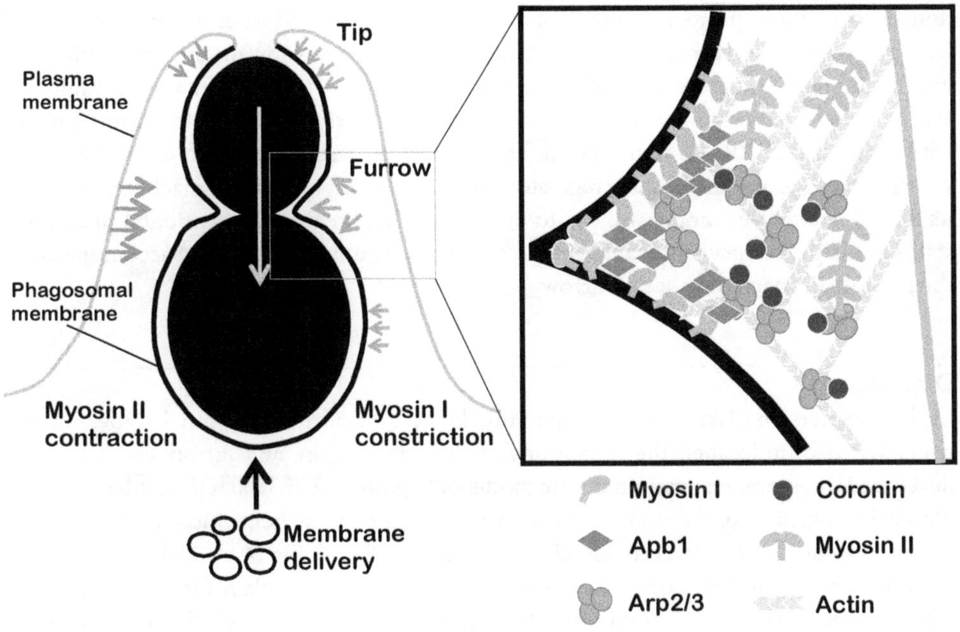

Figure 6.9 Overview of the phagocytic cup and molecular detail of the phagocytic furrow.

This study is complemented and extended by another recent study (Clarke *et al.*, 2010a) using a similar system. The authors reported the severing of budded yeasts at the neck by *Dictyostelium* cells (Fig. 6.9). Severing the neck of budded yeasts was still possible in myosin II-null cells, demonstrating that myosin II-dependent contraction is not responsible for the purse-like constriction. In line with this result, the authors show that cortexilin, a protein crucial for the maintenance of cortical tension is depleted from the phagocytic cup as is Myosin II.

Together these results highlight the existence of two contractile systems (Fig. 6.9). One cortical actin/myosin II-dependent system that drives ingression (pulling) of the plasma membrane in a way similar to cytokinesis. A second class I myosin-dependent system drives 'strangulation' (pushing) of the phagosomal membrane towards the particle in a way similar to the advance of a lamellipodium. These two concentric contractile systems allow for close apposition of the cup to the particle and ensure progression of the machinery on the particle surface consistent with a zipper mechanism.

In another study using FRAP with cells expressing GFP-actin, the authors (Clarke *et al.*, 2010a) showed that actin, not only at the phagocytic furrow, but all around the phagocytic cup is turned over in a matter of seconds (25 seconds for complete recovery of fluorescence). This again emphasizes how dynamic the F-actin network is around the phagocytic cup – a fact that is often underestimated. The authors then address the question of signalling at the furrow. They used probes for several phosphoinositides and showed that neither PI(3,4,5)P_3, PI(4,5)P_2 nor PI(3)P are enriched at the furrow. PI(3,4,5)P_3 appeared localized homogeneously around the phagocytic cup, PI(4,5)P_2 was restricted to the plasma membrane and not present in the cup and PI(3)P was detected only after cup closure and actin depolymerization. Monitoring of GFP tagged domains that bind activated Ras also showed no specific enrichment at the furrow. They then looked for the presence of an I-BAR domain-containing protein. These proteins have been involved in curvature sensing and have been shown to recruit members of the actin polymerization machinery to places of high membrane curvature through their SRC homology 3 (SH3) domains. A slight but maybe significant enrichment at the furrow of the only *Dictyostelium* protein containing an I-BAR domain (IBARa) was visualized. The authors suggest that this protein acts as a curvature-sensing probe and recruits regulators of the actin polymerization machinery through its SH3 domain. This model still has to be investigated further. As mentioned previously, stalling at negative curvature could still be a significant contribution to the accumulation of the phagocytic machinery at the furrow.

Discussion

In this chapter, we discussed phagocytic uptake in light of the conceptual zipper mechanism. We also underlined the limitations of this mechanism, and introduced a recently developed three-dimensional stochastic model of engulfment, in which the underlying biophysical mechanism of the zipper is an actin-driven thermal ratchet, which renders random membrane fluctuations irreversible close to the particle (Tollis *et al.*, 2010). Previously, such 'Brownian' ratchet models were successfully applied to explain force generation by actin polymerization and the motility of the pathogen *Listeria* in hosts cells (Mogilner *et al.*, 1996). For the first time, stochastic simulations can address the variability in particle uptake, cup growth (Van Zon *et al.*, 2009) and completion of endocytosis (Snijder *et al.*, 2009), and therefore address the question of bistability and cellular decision-making. As an example, in the model described in (Tollis *et al.*, 2010) a slight variation in some of the parameters leads either to complete or significantly incomplete uptake.

The development of such three-dimensional models of phagocytosis, which combine biophysics with signalling pathways, are now required to explain the strong dependence of phagocytosis on particle-shape. Experiments show that elongated particles are rapidly engulfed if attached to the cell membrane by the tip, but not if attached along its major axis (Champion and Mitragotri, 2006, 2009). Furthermore, spiral-shaped particles are not phagocytosed. The influence of particle-shape on phagocytosis is probably of biological relevance. On the one hand, it may increase the rate of infection of host cells by pathogenic bacteria. Indeed, recent experiments show that *Mycobacterium tuberculosis* and *Mycobacterium marinum* are efficiently taken up but later released tip first to disseminate to other tissues and spread the infection (Hagedorn *et al.*, 2009). On the other hand, the highly curved shapes of some bacteria, for instance spiral-shaped *Helicobacter* and *Campylobacter* species, may prevent their uptake by macrophages (Champion and Mitragotri, 2009), although injection of effector proteins can also be used by pathogens to hijack the immune or host cell's phagocytic response (Cornelis, 2002; Rosenberger and Finlay, 2003). Furthermore, the particle shape may be exploited to improve drug delivery by enclosing active drugs in particles, whose shape prevents uptake and destruction by macrophages (Champion and Mitragotri, 2009; Muttil *et al.*, 2009).

Research on phagocytosis and host–pathogen interactions has significantly advanced owing to a very good model organism: the lower eukaryote *Dictyostelium*. In a relatively small number of years, a wealth of molecular and genetic data has been gathered. *Dictyostelium* shares the evolutionary conserved basic mechanistic machinery such as actin, myosins, and small GTPases with human macrophages. Both *Dictyostelium* cells and macrophages are potential hosts for many invasive bacteria, which exploit phagocytosis to penetrate the cells (Cosson and Soldati, 2008). Owing to manageable genetics and cell culture and its status as a professional phagocyte, it has been an excellent model organism for live-cell imaging. Phenomena such as phospholipid microdomains, actin waves and differential myosin localization were successfully imaged. These data need to be integrated into new predictive models of phagocytosis and then applied to host–pathogen interactions.

Beyond phagocytosis, zipper-like mechanisms are relevant in a variety of other biological situations. One such example is sporulation during starvation in *Bacillus subtilis*. After asymmetric cell division, the larger mother cell engulfs the smaller forespore for spore maturation. Interestingly, the mother cell even engulfs the forespore when the cell wall is artificially removed. This process occurs in a fast, zipper-like fashion without known sources of energy (Broder and Pogliano, 2006). Importantly, forespore engulfment is subject to high variation. About 60% of the cells successfully complete forespore engulfment, while 40% do not, similar to the observation of the mechanical bottleneck in phagocytosis. Other examples of engulfment may not be driven by a zipper mechanism. For instance, the penetration of red-blood cells by the malaria *Plasmodium* merozoite is driven by an elaborate parasite actin machinery), which devotes all its resources to wrap the cell membrane around itself. In stark contrast to phagocytosis, the engulfing host cell is completely passive (Cowman and Crabb, 2006).

Zippering and ratchet mechanisms can also occur at a multicellular scale, for example during dorsal closure in the *Drosophila* fruit fly. Halfway through embryonic development, the epidermis exhibits a gap at the dorsal side covered by an extra-embryonic epithelium, the amnioserosa. Dorsal closure is the process whereby interactions between the two epithelia establish epidermal continuity, which proceeds in a zipper-like fashion (Gorfinkiel *et al.*, 2009). First, a supracellular contractile actin cable is generated that surrounds the gap for closure. Then, through force generation by contraction of the cells from the amnioserosa, dorsal closure occurs in two phases: (1) a slow phase during which apical contraction of cells at the posterior end predominates over contraction of centrally located cells; (2) a subsequent fast phase, during which all the cells engage in the contraction, which correlates with the zippering process. Therefore, the force produced by amnioserosa cells is only transient, pulsed, and occurs long before dorsal closure is actually observable. However, the presence of the actin cable prevents counteracting ventral-ward epidermis relaxation after the force pulse, acting like a ratchet, which ultimately leads to the closure of the central region well after the end of the force pulse. As a consequence, this work uncovered a sophisticated mechanism of cooperative force generation between two major forces driving morphogenesis (Solon *et al.*, 2009).

Finally, deciphering the mechanisms of phagocytosis is also extremely relevant in host–pathogen interactions [see Pizarro-Cerda and Cossart (2006) and references therein]. Several pathogens seem to have exploited their interaction with host-cell receptors to either adhere or even trigger actin cytoskeleton rearrangements of the host cell that can lead to cellular invasion. Enteropathogenic bacteria such as *Yersinia enterocolitica* and *Yersinia pseudotuberculosis* use surface proteins such as invasin (InvA), to bind with high affinity to host cell's integrins, mimicking normal interactions of these adhesion molecules with a substrate. Integrin engagement with InvA leads also to activation of several small GTPases, including Rac1, an actin regulator, and Arf6, which has been implicated in membrane trafficking. Downstream, $PI(3,4)P_2$ is produced at the bacterial entry site, activating and localizing a variety of actin-regulating molecules, ultimately leading to the cell wrapping its membrane around the bacterium leading to infection of the host cell by *Yersinia*. This invasive behaviour is only required to penetrate through epithelial layers. Later in the bacterial infection, *in vivo* proliferation occurs in the extracellular space. This relies on an important arsenal of anti-phagocytic molecules developed by the pathogenic *Yersinia* species in order to escape destruction by macrophages.

Acknowledgements

R.G.E. acknowledges funding from the Biotechnology and Biological Sciences Research Council grant BB/G000131/1 and from the Centre for Integrative Systems Biology at Imperial College (CISBIC). We thank the members of the Soldati lab for stimulating discussions and the Swiss National Science Foundation for funding.

References

Ackerman, A.L., Giodini, A., and Cresswell, P. (2006). A role for the endoplasmic reticulum protein retrotranslocation machinery during cross presentation by dendritic cells. Immunity 25, 607–617.

Aderem, A., and Underhill, D.M. (1999). Mechanisms of phagocytosis in macrophages. Annu. Rev. Immunol. *17*, 593–623.

Allen, L.A., and Aderem, A. (1996). Molecular definition of distinct cytoskeletal structures involved in complement- and Fc receptor-mediated phagocytosis in macrophages. J. Exp. Med. *184*, 627–637.

Ananthakrishnan, R., and Ehrlicher, A. (2007). The forces behind cell movement. Int. J. Biol. Sci. 3, 303–317.

Andrews, N.L., Lidke, K.A., Pfeiffer, J.R., Burns, A.R., Wilson, B.S., Oliver, J.M., and Lidke, D.S. (2008). Actin restricts FczRI diffusion and facilitates antigen-induced receptor immobilization. Nat. Cell Biol. *10*, 955–963.

Araki, N., Johnson, M.T., and Swanson, J.A. (1996). A role for phosphoinositide 3-kinase in the completion of macropinocytosis and phagocytosis by macrophages. J. Cell Biol. *135*, 1249–1260.

Araki, N., Hatae, T., Furukawa, A., and Swanson, J.A. (2003). Phosphoinositide-3-kinase-independent contractile activities associated with Fcgamma-receptor-mediated phagocytosis and macropinocytosis in macrophages. J. Cell. Sci. *116*, 247–257

Arora, P.D., Chan, M.W., Anderson, R.A., Janmey, P.A., and McCulloch, C.A. (2005). Separate functions of gelsolin mediate sequential steps of collagen phagocytosis. Mol. Biol. Cell *16*, 5175–5190.

Bajno, L., Peng, X.R., Schreiber, A.D., Moore, H.P., Trimble, W.S., and Grinstein, S. (2000). Focal exocytosis of VAMP3-containing vesicles at sites of phagosome formation. J. Cell Biol. *149*, 697–705.

Becker, T., Volchnuk, A., and Rothman, J.E. (2004). Differential use of endoplasmic reticulum membrane for phagocytosis in J774 macrophages. Proc. Natl. Acad. Sci. U.S.A. *102*, 4022–4026.

Benabentos, R., Hirose, S., Sucgang, R., Curk, T., Katoh, M., Ostrowski, E.A., Strassmann, J.E., Queller, D.C., Zupan, B., Shaulsky, G., *et al.* (2009). Polymorphic members of the lag gene family mediate kin discrimination in *Dictyostelium*. Curr. Biol. *19*, 567–572.

Beningo, K.A., and Wang, Y.L. (2002). Fc-receptor-mediated phagocytosis is regulated by mechanical properties of the target. J. Cell Sci. *115*, 849–856.

Berger, M., O'Shea, J., Cross, A.S., Folks, T.M., Chused, T.M., Brown, E.J., and Frank, M.M. (1984). Human neutrophils increase expression of C3bi as well as C3b receptors upon activation. J. Clin. Invest. *74*, 1566–1571.

Bivas, I., Georgescauld, D., Jeandaine, N., Winterhalter, M., Méléard, P., Marinov, G., and Bothorel, P. (1997). Elastic properties of lipid bilayer containing modified lipids. Prog. Coll. Pol. Sci. *105*, 197–203.

Blystone, S.D., and Brown, E.J. (1999). Integrin receptors of phagocytes. Adv. Cell. Mol. Biol. Memb. Organ. *5*, 103–147.

Booth, J.W., Trimble, W.S., and Grinstein, S. (2001). Membrane dynamics in phagocytosis. Semin. Immunol. *13*, 357–364.

Bohdanowicz, M., Cosío, G., Backer, J.M., and Grinstein, S. (2010). Class I and class III phosphoinositide 3-kinases are required for actin polymerization that propels phagosomes. J. Cell Biol. *191*, 999–1012.

Boulais, J., Trost, M., Landry, C.R., Dieckmann, R., Levy, E.D., Soldati, T., Michnick, S.W., Thibault, P., and Desjardins, M. (2010). Molecular characterization of the evolution of phagosomes. Mol. Syst. Biol. *6*, 423.

Bretschneider, T., Anderson, K., Ecke, M., Müller-Taubenberger, A., Schroth-Diez, B., Ishikawa-Ankerhold, H.C., and Gerisch, G. (2009). The three-dimensional dynamics of actin waves, a model of cytoskeletal self-organization. Biophys. J. *96*, 2888–2900.

Broder, D., and Pogliano, K. (2006). Forespore engulfment mediated by a ratchet-like mechanism. Cell *126*, 917–928.

Cai, L., Makhov, A.M., Schafer, D.A., and Bear, J.E. (2008). Coronin 1B antagonizes cortactin and remodels Arp2/3-containing actin branches in lamellipodia. Cell *134*, 828–842.

Cannon, G.J., and Swanson, J.A. (1992). The macrophage capacity for phagocytosis. J. Cell Sci. *101*, 907–913.

Caron, E., and Hall, A. (1998). Identification of two distinct mechanisms of phagocytosis controlled by different Rho GTPases. Science *282*, 1717.

Caron, E., Self, A.J., and Hall, A. (2000). The GTPase Rap1 controls functional activation of macrophage integrin alphaMbeta2 by LPS and other inflammatory mediators. Curr. Biol. *10*, 974–978.

Castellano, F., Chavrier, P., and Caron, E. (2001). Actin dynamics during phagocytosis. Semin. Immunol. *13*, 347–355.

Champion, J.A., and Mitragotri, S. (2006). Role of target geometry in phagocytosis. Proc. Natl. Acad. Sci. U.S.A. *103*, 4930–4934.

Champion, J.A., and Mitragotri, S. (2009). Shape induced inhibition of phagocytosis of polymer particles. Pharm. Res. *26*, 244–249.

Chan, C., Beltzner, C.C., and Pollard, T.D. (2009). Cofilin dissociates Arp2/3 complex and branches from actin filaments. Curr. Biol. *19*, 537–545.

Charras, G.T., Hu, C.K., Coughlin, M., and Mitchison, T.J. (2006). Reassembly of contractile actin cortex in cell blebs. J. Cell Biol. *175*, 477–490.

Chen, G., Zhuchenko, O., and Kuspa A. (2007). Immune-like phagocyte activity in the social amoeba. Science *317*, 678–681.

Clarke, M., Kohler, J., Arana, Q., Liu, T., Heuser, J., and Gerisch, G. (2002). Dynamics of the vacuolar H(+)-ATPase in the contractile vacuole complex and the endosomal pathway of *Dictyostelium* cells. J. Cell Sci. *115*, 2893–2905.
Clarke, M., Engel, U., Giorgione, J., Muller-Taubenberger, A., Prassler, J., Veltman, D., and Gerisch, G. (2010a). Curvature recognition and force generation in phagocytosis. BMC Biol. *8*, 154.
Clarke, M., Maddera, L., Engel, U., and Gerisch, G. (2010b). Retrieval of the vacuolar H-ATPase from phagosomes revealed by live cell imaging. PLoS One *5*, e8585.
Colombelli, J., Besser, A., Kress, H., Reynaud, E.G. Girard, P., Caussinus, E., Haselmann, U., Small, J.V., Schwarz, U.S., and Stelzer, E.H.K. (2009). Mechanosensing in actin stress fibers revealed by a close correlation between force and protein localization. J. Cell Sci. *122*, 1665–1679.
Conner, S.D., and Schmid, S.L. (2008). Regulated portals of entry into the cell. Nature *422*, 37–44.
Corbett-Nelson, E.F., Mason, D., Marshall, J.G., Collette, Y., and Grinstein, S. (2006). Signaling-dependent immobilization of acylated proteins in the inner monolayer of the plasma membrane. J. Cell Biol. *174*, 255–265.
Cornelis, G.R. (2002). The Yersinia Ysc-Yop 'Type III' weaponry. Nat. Rev. Mol. Cell Biol. *3*, 742–754.
Cornillon, S., Gebbie, L., Benghezal, M., Nair, P., Keller, S., Wehrle-Haller, B., Charette, S.J., Bruckert, F., Letourneur, F., and Cosson, P. (2006). An adhesion molecule in free-living *Dictyostelium* amoebae with integrin beta features. EMBO Rep. *7*, 617–621.
Cornillon, S., Froquet, R., and Cosson, P. (2008). Involvement of Sib proteins in the regulation of cellular adhesion in *Dictyostelium discoideum*. Eukaryot. Cell *7*, 1600–1605.
Cosson, P., and Soldati, T. (2008). Eat, kill or die: when amoeba meets bacteria. Curr. Opin. Microbiol. *11*, 271–276.
Cowman, A.F., and Crabb, B.S. (2006). Invasion of red blood cells by malaria parasites. Cell *124*, 755–766.
Cox, D., Tseng, C.C., Bjekic, G., and Greenberg, S. (1999). A requirement for phosphatidylinositol 3-kinase in pseudopod extension. J. Biol. Chem. *274*, 1240–1247.
Cox, D., Berg, J.S., Cammer, M., Chinegwundoh, J.O., Dale, B.M., Cheney, R.E., and Greenberg, S. (2002). Myosin X is a downstream effector of PI(3)K during phagocytosis. Nature Cell Biol. *4*, 469–477.
Daëron, M. (1997) Fc receptor biology. Ann. Rev. Immunol. *15*, 203–234.
Decuzzi, P., and Ferrari, M. (2007). The role of non-specific interactions in receptor-mediated endocytosis of nanoparticles. Biomaterials *28*, 2915–2922.
Decuzzi, P., and Ferrari, M. (2008) The receptor-mediated endocytosis of nonspherical particles. Biophys. J. *94*, 3790–3797.
Desjardins, M. (2003). ER-mediated phagocytosis: a new membrane for new functions. Nat. Rev. Immunol. *3*, 280–291.
Desjardins, M., Huber, L.A., Parton, R.G., and Griffiths, G. (1994). Biogenesis of phagolysosomes proceeds through a sequential series of interactions with the endocytic apparatus. J. Cell Biol. *124*, 677–688.
Di, A., Nelson, D.J., Bindokas, V., Brown, M.E., Libunao, F., and Palfrey, H.C. (2003). Dynamin regulates focal exocytosis in phagocytosing macrophages. Mol. Biol. Cell *14*, 2016–2028.
Dieckmann, R., von Heyden, Y., Kistler, C., Gopaldass, N., Hausherr, S., Crawley, S.W., Schwarz, E.C., Diensthuber, R.P., Cote, G.P., Tsiavaliaris, G., *et al.* (2010). A myosin IK-Abp1-PakB circuit acts as a switch to regulate phagocytosis efficiency. Mol. Biol. Cell *21*, 1505–1518.
Dieckmann, R., Monroy, R., Ruppert, T., Bloomfield, G., and Soldati, T. (2011). A balance in the delivery of ER components and the vacuolar proton pump determines phagosome acidification. (Submitted.)
Dormann, D., Weijer, G., Dowler, S., and Weijer, C.J. (2004). *In vivo* analysis of 3-phosphoinositide dynamics during *Dictyostelium* phagocytosis and chemotaxis. J. Cell. Sci. *117*, 6497–6509.
Dupuy, A.G., and Caron, E. (2008). Integrin-dependent phagocytosis: spreading from microadhesion to new concepts. J. Cell. Sci. *121*, 1773–1783.
Ehlenberger, A.G., and Nussenzweig, V. (1977). The role of membrane receptors for C3b and C3d in phagocytosis. J. Exp. Med. *145*, 357–71.
Flannagan, R.S., Harrison, R.E., Jagaman, K., and Grinstein, S. (2010). Dynamic macrophage "probing" is required for the efficient capture of phagocytic targets. J. Cell Biol. *191*, 1205–1218.
Francis, C.L., Starnbach, M.N., and Falkow, S. (1992). Morphological and cytoskeletal changes in epithelial cells occur immediately upon interaction with Salmonella typhimurium grown under low-oxygen conditions. Mol. Microbiol. *6*, 3077–3087.
Francis, C.L., Ryan, T.A., Jones, B.D., Smith, S.J., and Falkow, S. (1993). Ruffles induced by *Salmonella* and other stimuli direct macropinocytosis of bacteria. Nature *364*, 639–642.

Gagnon, E., Duclos, S., Rondeau, C., Chevet, E., Cameron, P.H., Steele-Mortimer, O., Paiement, J.,Bergeron, J.J., and Desjardins, M. (2002). Endoplasmic-reticulum-mediated phagocytosis is a mechanism of entry into macrophages. Cell *110*, 119–131.

Galan, J.E., Pace, J., and Hayman, M.J. (1992). Involvement of the epidermal growth factor receptor in the invasion of cultured mammalian cells by Salmonella typhimurium. Nature *357*, 588–589.

Gao, H., Shi, W., and Freund, L.B. (2005). Mechanics of receptor-mediated endocytosis. Proc. Natl. Acad. Sci. U.S.A. *102*, 9464–9474.

Gerisch, G., Ecke, M., Schroth-Diez, B., Gerwig, S., Engel, U., Maddera, L., and Clarke, M. (2009). Self-organizing actin waves as planar phagocytic cup structures. Cell Adh. Migr. *3*, 373–382.

Giorgione, J., and Clarke, M. (2008). Heterogeneous modes of uptake for latex beads revealed through live cell imaging of phagocytes expressing a probe for phosphatidylinositol-(3,4,5)-trisphosphate and phosphatidylinositol-(3,4)-bisphosphate. Cell Motil. Cytoskeleton *65*, 721–733.

Gold, E.S., Underhill, D.M., Morrissette, N.S., Guo, J., McNiven, M.A., and Aderem, A. (1999). Dynamin 2 is required for phagocytosis in macrophages. J. Exp. Med. *190*, 1849–1856.

Goldstein, B., Faeder, J.R., Hlavacek, W.S., Blinov, M.L., Redondo, A., and Wofsy, C. (2001). Modeling of early signalling events mediated by FcεRI. Mol. Immunol. *38*, 1213–1219.

Goodall, R.J., Lai, Y.F., and Thompson, J.E. (1972). Turnover of plasma membrane during phagocytosis. J. Cell. Sci. *3*, 525.

Gorfinkiel, N,. Blanchard, G.B., Adams, R.J., Martinez, and Arias, A. (2009). Mechanical control of global cell behaviour during dorsal closure in *Drosophila*. Development *136*, 1889–1898.

Greenberg, S., and Grinstein, S. (2002). Phagocytosis and innate immunity. Curr. Opin. Immunol. *14*, 136–145.

Griffin, F.M., Griffin, J.A., Leider, J.E., and Silverstein, S.C. (1975). Studies on the mechanism of phagocytosis. J. Exp. Med. *142*, 1263–1282.

Griffin, F.M., Griffin, J.A., Leider, J.E., and Silverstein, S.C.(1976). Studies on the mechanism of phagocytosis. II. The interaction of macrophages with anti-immunoglobulin IgG-coated bone marrow-derived lymphocytes. J. Exp. Med. *144*, 788–809.

Grinstein, S. (2010). Imaging signal transduction during phagocytosis: phospholipids, surface charge, and electrostatic interactions. Am. J. Physiol. Cell Physiol. *299*, C876–881.

Groves, A., Dart, A.E., Covarelli, V., and Caron, E. (2008). Molecular mechanisms of phagocytic uptake in mammalian cells. Cell. Mol. Life Sci. *65*, 1957–1976.

Guetta, D., Langou, K., Grunwald, D., Klein, G., and Aubry, L. (2010). FYVE-dependent endosomal targeting of an arrestin-related protein in amoeba. PLoS One *5*, e15249.

Hagedorn, M., Rohde, K.H., Russell, D.G., and Soldati, T. (2009). Infection by tubercular mycobacteria is spread by nonlytic ejection from their amoeba hosts. Science *323*, 1729–1733.

Hallett, M.B., and Dewitt, S. (2007). Ironing out the wrinkles of neutrophil phagocytosis. Tr. Cell Biol. *17*, 209–214.

Hampton, M.B., Kettle, A.J., and Winterbourn, C.C. (1998). Inside the neutrophil phagosome: oxidants, myeloperoxidase, and bacterial killing. Blood *92*, 3007–3017.

Hao, J.J., Liu, Y., Kruhlak, M., Debell, K.E., Rellahan, B.L., and Shaw, S. (2009). Phospholipase C-mediated hydrolysis of PIP2 releases ERM proteins from lymphocyte membrane. J. Cell Biol. *84*, 451–462.

Helfrich, W. (1973). Elastic properties of lipid bilayers: theory and possible experiments. Z. Naturforsch C. *28*, 693–703.

Herant, M., Marganski, W.A., and Dembo, M. (2003). The mechanics of neutrophils: synthetic modeling of three experiments. Biophys. J. *84*, 3389–3413.

Herant, M., Heinrich, V., and Dembo, M. (2005). Mechanics of neutrophils phagocytosis: behavior of the cortical tension. J. Cell Sci. *118*, 1789–1797.

Herant, M., Heinrich, V., and Dembo, M. (2006). Mechanics of neutrophils phagocytosis: experiments and quantitative models. J. Cell Sci. *119*, 1903–1913.

Holevinski, K.O., and Nelson, D.J. (1998). Membrane capacitance changes associated with particle uptake during phagocytosis in macrophages. Biophys. J. *75*, 2577–2586.

Hong, J., Kim, J., Baek, J., Cha, M., Lee, J., and Park, S. (2007). Cell Migration Driven by a Mechanical Stiffness Gradient. Proceedings of the second IEEE International Conference on Nano/Micro Engineered and Molecular Systems (NEMS), pp. 797–800.

Hynes, R.O. (1992). Integrins: versatility, modulation, and signalling in cell adhesion. Cell *69*, 11–25.

Isberg, R.R., and VanNhieu, G.T. (1995). The mechanism of phagocytic uptake promoted by invasin-integrin interaction. Tr. Cell Biol. *5*, 120–124.

Jaumouille, V., and Grinstein, S. (2010) Receptor mobility, the cytoskeleton, and particle binding during phagocytosis. Curr. Opin. Cell Biol. *23*, 1–8.
Ji, L., Lim, J., and Danuser, G. (2008). Fluctuations of intracellular forces during cell protrusion. Nat. Cell Biol. *10*, 1393.
Jeon, K.W., and Jeon, M.S. (1976). Scanning electron microscope observations of *Amoeba proteus* during phagocytosis. J. Euk. Microbiol. *23*, 83–86.
Jones, S.L., Knaus, U.G., Bokoch, G.M., and Brown, E.J. (1998). Two signalling mechanisms for activation of alphaM beta2 avidity in polymorphonuclear neutrophils. J. Biol. Chem. *273*, 10556–66.
Jung, G., Remmert, K., Wu, X., Volosky, J.M., and Hammer, J.A. 3rd. (2001). The *Dictyostelium* CARMIL protein links capping protein and the Arp2/3 complex to type I myosins through their SH3 domains. J. Cell Biol. *153*, 1479–1497.
Kaplan, G. (1977). Differences in the mode of phagocytosis with Fc and C3 receptors in macrophages. Scan. J. Immunol. *6*, 797–807.
Kastner, P.M., Schleicher, M., and Muller-Taubenberger, A. (2010). The NDR family kinase NdrA of *Dictyostelium* localizes to the centrosome and is required for efficient phagocytosis. Traffic *12*, 301–312.
Kinchen, J.M., and Ravichandran, K.S. (2008). Phagosome maturation: going though the acid test. Nat. Rev. Mol. Cell Biol. *9*, 781–795.
Kraft, S., and Kinet, J.P. (2007). New developments in FcεRI regulation, function and inhibition. Nat. Rev. Immunol. *7*, 365–378.
Kress, H., Stelzer, E.H.K., Holzer, D., Buss, F., Griffiths, G., and Rohrbach, A. (2007). Filopodia act as phagocytic tentacles and pull with discrete steps and a load-dependent velocity. Proc. Natl. Acad. Sci. U.S.A. *104*, 11633–11638.
Lai-king, N.G., Sherburne, R., Taylor, D.E., and Stiles, M.E. (1985). Morphological forms and viability of *Campylobacter* species studied by electron microscopy. J. Bacteriol. *164*, 338–343.
Le Cabec, V., Carreno, S., Mois, A., Bordier, C., and Maridonneau-Parini, I. (2002). Complement receptor 3 (CD11b/CD18) mediates type I and type II phagocytosis during nonopsonic and opsonic phagocytosis, respectively. J. Immunol. *169*, 2003–2009.
Lee, W.L., Mason, D., Schreiber, A.D., and Grinstein, S. (2007). Quantitative analysis of membrane remodelling at the phagocytic cup. Mol. Biol. Cell *18*, 2883–2892.
Lo, C.M., Wang, H.B., Dembo, M., and Wang, Y.L. (2000). Cell movement is guided by the rigidity of the substrate. Biophys. J. *79*, 144–152.
Luo, Y., Isaac, B.M., Casadevall, A., and Cox, D. (2009). Phagocytosis inhibits F-actin-enriched membrane protrusions stimulated by fractalkine (CX3CL1) and colony-stimulating factor 1. Infect. Immun. *77*, 4487–4495.
Mahadeo, D.C., and Parent, C.A. (2006). Signal relay during the life cycle of *Dictyostelium*. Curr. Top. Dev. Biol. *73*, 115–140.
May, R.C., and Machesky, L.M. (2001). Phagocytosis and the actin cytoskeleton. J. Cell. Sci. *114*, 1061–1077.
Medalia, O., Beck, M., Ecke, M., Weber, I., Neujahr, R., Baumeister, W., and Gerisch, G. (2007). Organization of actin networks in intact filopodia. Curr. Biol. *17*, 79–84.
Mercanti, V., Charette, S.J., Bennett, N., Ryckewaert, J.J., Letourneur, F., and Cosson, P. (2006). Selective membrane exclusion in phagocytic and macropinocytic cups. J. Cell Sci. *119*, 4079–4087.
Mogilner, A., and Oster, G. (1996). Cell motility driven by actin polymerization. Biophys. J. *71*, 3030–3045.
Mogilner, A., and Oster, G. (2003). Force generation by actin polymerization II; the elastic ratchet and tethered filaments. Biophys. J. *84*, 1591–1605.
Muttil, P., Wang, C., and Hickey, A.J. (2009) Inhaled drug delivery for tuberculosis therapy. Pharm. Res. *26*, 2401–2416.
Muller-Taubenberger, A., Lupas, A.N., Li, H., Ecke, M., Simmeth, E., and Gerisch, G. (2001). Calreticulin and calnexin in the endoplasmic reticulum are important for phagocytosis. EMBO J. *20*, 6772–6782.
Neuhaus, E.M., and Soldati, T. (1999). Molecular mechanisms of membrane trafficking. What do we learn from *Dictyostelium discoideum*? Protist *150*, 235–243.
Neuhaus, E.M., Almers, W., and Soldati, T. (2002). Morphology and dynamics of the endocytic pathway in *Dictyostelium discoideum*. Mol. Biol. Cell *13*, 1390–1407.
Niedergang, F., and Chavrier, P. (2004). Signaling and membrane dynamics during phagocytosis: many roads lead to the phagos(R)ome. Curr. Opin. Cell Biol. *16*, 422–428.
Niedergang, F., Colucci-Guyon, E., Dubois, T., Raposo, G., and Chavrier, P. (2003). ADP ribosylation factor 6 is activated and controls membrane delivery during phagocytosis in macrophages. J. Cell Biol. *161*, 1143–1150.

Peyron, P., Maridonneau-Parini, I., and Stegmann, T. (2001). Fusion of human neutrophil phagosomes with lysosomes *in vitro*: involvement of tyrosine kinases of the Src family and inhibition by mycobacteria. J. Biol. Chem. *276*, 35512–35517.

Pizarro-Cerda, J., and Cossart, P. (2006). Bacterial adhesion and entry into host cells. Cell *124*, 715.

Racoosin, E.L., and Swanson, J.A. (1989). Macrophage colony-stimulating factor (rM-CSF) stimulates pinocytosis in bone marrow-derived macrophages. J. Exp. Med. *170*, 1635–1648.

Ravetch, J.V., and Bolland, S. (2001). IgG Fc receptors. Annu. Rev. Immunol. *19*, 275–290.

Rosenberger, C.M., and Finlay, B.B. (2003). Phagocyte sabotage: disruption of macrophage signalling by bacterial pathogens. Nat. Rev. Mol. Cell Biol. *4*, 385–396.

Segal, A.W. (2005). How neutrophils kill microbes. Ann. Rev. Immunol. *23*, 197–223.

Sengelov, H., Kjeldsen, L., Diamond, M.S., Springer, T.A., and Borregaard, N. (1993). Subcellular localization and dynamics of Mac-1 (alpha m beta 2) in human neutrophils. J. Clin. Invest. *92*, 1467–76.

Serrander, L., Skarman, P., Rasmussen, B., Witke, W., Lew, D.P., Krause, K.H., Stendhal, O., and Nusse, O. (2000). Selective inhibition of IgG-mediated phagocytosis in gelsoline-deficient murine neutrophils. J. Immunol. *165*, 2451–2457.

Silva, E., Au-Yeung, H.W., Van Goethem, E., Burden, J., and Franc, N.C. (2007). Requirement for a *Drosophila* E3-ubiquitin ligase in phagocytosis of apoptotic cells. Immunity *27*, 585–596.

Snijder, B., Sacher, R., Rämö, P., Damm, E.M., Liberali, P., and Pelkmans, L. (2009). Population context determines cell-to-cell variability in endocytosis and virus infection. Nature *461*, 520–523.

Sobota, A., Strzelecka-Kiliszek, A., Gladkowska, A., Yoshida, K., Mrozinska, K., and Kwiatkowska, K. (2005). Binding of IgG-opsonized particles to FcγR is an active stage of phagocytosis that involves receptor clustering and phosphorylation. J. Immunol. *175*, 4450–4457.

Soldati, T. (2003). Unconventional myosins, actin dynamics and endocytosis: a menage a trois? Traffic *4*, 358–366.

Soldati, T., and Schliwa, M. (2006). Powering membrane traffic in endocytosis and recycling. Nat. Rev. Mol. Cell Biol. *7*, 897–908.

Solon, J., Kaya-Copur, A., Colombelli, J., and Brunner, D. (2009). Pulsed forces timed by a ratchet-like mechanism drive directed tissue movement during dorsal closure. Cell *137*, 1331–1342.

Svitkina, T., and Borisy, G.G. (1999). Arp2/3 complex and actin depolymerizing factor/cofilin in dendritic organization and treadmilling of actin filament array in lamellipodia. J. Cell Biol. *145*, 1009–1026.

Swanson, J.A. (2008). Shaping cups into phagosomes and macropinosomes. Nat. Rev. Mol. Cell. Biol. *9*, 639–649.

Swanson, J.A., and Hoppe, A.D. (2004). The coordination of signaling during Fc receptor-mediated phagocytosis. J. Leukoc. Biol. *76*, 1093–1103.

Swanson, J.A., Johnson, M.T., Beningo, K., Post, P., Mooseker, M., and Araki, N. (1999). A contractile activity that closes phagosomes in macrophages. J. Cell Sci. *112*, 307–316.

Takeuchi, A. (1967). Electron microscope studies of experimental *Salmonella* infection. I. Penetration into the intestinal epithelium by *Salmonella typhimurium*. Am. J. Pathol. *50*, 109–120.

Thompson, C.R.L., and Bretscher, M.S. (2002). Cell polarity and locomotion, as well as endocytosis, depend on NSF. Development *129*, 4185–4192.

Tohyama, Y., and Yamamura, H. (2008). Complement-mediated phagocytosis: the role of Syk. IUBMB life *58*, 304–308.

Tollis, S., Dart, A.E., Tzircotis, G., and Endres, R.G. (2010). The zipper mechanism in phagocytosis: energetic requirements and variability in phagocytic cup shape. BMC Sys. Biol. *4*, 149.

Underhill, D.M., and Ozinsky, A. (2002). Phagocytosis of microbes: complexity in action. Ann. Rev. Immunol. *20* 825–852.

Van Zon, J.S., Tzircotis, G., Caron, E., and Howard, M. (2009). A mechanical bottleneck explains the variation in cup growth during FcγR phagocytosis. Mol. Syst. Biol. *5*, 298.

Wienke, D.C., Knetsch, M.L., Neuhaus, E.M., Reedy, M.C., and Manstein, D.J. (1999). Disruption of a dynamin homologue affects endocytosis, organelle morphology, and cytokinesis in *Dictyostelium discoideum*. Mol. Biol. Cell *10*, 225–243.

Wright, S.D., and Detmers, P.A. (1988). Adhesion-promoting receptors on phagocytes. J. Cell Sci. *9*, 99–120.

Wright, S.D., and Griffin, F.M., Jr. (1985) Activation of phagocytic cells' C3 receptors for phagocytosis. J. Leukoc. Biol. *38*, 327–39.

Wright, S.D., and Silverstein, S.C. (1982). Tumor-promoting phorbol esters stimulate C3b and C3b' receptor-mediated phagocytosis in cultured human monocytes. J. Exp. Med. *156*, 1149–1164.

Yin, H.L., Albrecht, J.H., and Fattoum, A. (1981). Identification of gelsolin, a Ca^{2+}-dependent regulatory protein of actin gel–sol transformation, and its intracellular distribution in a variety of cells and tissues. J. Cell Biol. *91*, 901–906.
Yu, X., Lu, N., and Zhou, Z. (2008). Phagocytic receptor CED-1 initiates a signaling pathway for degrading engulfed apoptotic cells. PLoS Biol. *6*, e61.
Zanchi, R., Howard, G., Bretscher, M.S., and Kay, R.R. (2010). The exocytic gene secA is required for *Dictyostelium* cell motility and osmoregulation. J. Cell Sci. *123*, 3226–3234.
Zhang, Y., Hoppe, A.D., and Swanson, J.A. (2010). Coordination of Fc receptor signaling regulates cellular commitment to phagocytosis. Proc. Natl. Acad. Sci. U.S.A. *107*, 19332–19337.

System-level Strategies for Studying the Metabolism of *Mycobacterium tuberculosis*

7

Dany J.V. Beste and Johnjoe McFadden

Abstract

Despite decades of research many aspects of the biology of *Mycobacterium tuberculosis* remain unclear and this is reflected in the antiquated tools available to treat and prevent tuberculosis. Consequently, this disease remains a serious public health problem responsible for two to three million deaths each year. Important discoveries linking *M. tuberculosis* metabolism and pathogenesis have renewed interest in the metabolic underpinning of the interaction between the pathogen and its host. Whereas, previous experimental studies tended to focus on the role of single genes, antigens or enzymes the central paradigm of systems biology is that the role of any gene cannot be determined in isolation from its context. Therefore systems approaches examine the role of genes and proteins embedded within a network of interactions. We here examine the application of this approach to studying metabolism of *M. tuberculosis*. Recent advances in high-throughput experimental technologies, such as functional genomics and metabolomics, provide datasets that can be analysed with computational tools such as flux balance analysis. These new approaches allow metabolism to be studied on a genome scale and have already been applied to gain insights into the metabolic pathways utilized by *M. tuberculosis in vitro* and identify potential drug targets. The information from these studies will fundamentally change our approach to tuberculosis research and lead to new targets for therapeutic drugs and vaccines.

Introduction

Tuberculosis (TB) is a disease that plagued ancient Egyptians and still remains a major threat to human health thousands of years later. The control of tuberculosis has been significantly hindered by the limited resources available for both the prevention and treatment of tuberculosis. A truly effective vaccine is lacking as the 90-year-old *Mycobacterium bovis* bacillus Calmette–Guerin live attenuated vaccine is not universally protective and does not produce immunity against re-infection or reactivation. Lengthy (6–9 months) and complex (three or more different drugs) treatment is required using currently available anti-TB drugs. The economic and logistic burden of administering these drug regimens in industrially undeveloped countries where TB is most prevalent is enormous and combined with poor patient compliance are important factors in the emergence of drug-resistant TB isolates that are causing on-going epidemics. These factors underscore the urgent need for the development of novel and effective therapeutics and vaccines and new approaches will be required to achieve these goals.

Mycobacterium tuberculosis is an unusual bacterial pathogen, which has the remarkable ability to cause both acute life-threatening disease and also clinically latent infections which can persist for the lifetime of the human host. Unlike many pathogens *M. tuberculosis* does not rely on the production of specific toxins to cause disease but rather the secret of this bacterium's great success seems to be the ability to adapt and survive within the changing and adverse environment provided by the human host during the course of an infection. It is becoming apparent that key to this adaptation is the metabolic reprogramming of *M. tuberculosis* during both the acute and chronic phase of TB disease and therefore a more complete understanding of mycobacterial metabolism remains a major goal of TB research.

Whilst recent increases in research funding has progressed our understanding of the basic biology of *M. tuberculosis* this has not yet impacted on the global TB trends which remain at staggering levels. A possible reason why it has been difficult to translate basic research into effective strategies for combating tuberculosis is that TB research has until recently, focused on studying individual parameters in isolation which can consequently result in an overestimation of the importance of these factors. This effect may be particularly profound for a persistent pathogen such as *M. tuberculosis* that lacks classical virulence factors. The systems biology framework, which investigates the dynamic interactions of many components, provides an alternative and complementary strategy to the more traditional reductionist approaches to TB research. This methodology has started to be applied to the metabolism of *M. tuberculosis* on a genome scale and promises to drive biological discovery in the TB research field by providing scaffolding for the interpretation of 'omic' scale datasets, directing hypothesis driven discovery and also assisting in the identification of novel drug targets.

Metabolism of *M. tuberculosis*

Application of metabolic modelling approaches to *M. tuberculosis* is aided by the fact that metabolism is a reasonably well studied system even in mycobacteria. Moreover, metabolism has been shown to be involved in the virulence of *M. tuberculosis,* playing a key role in the development and maintenance of both acute and persistent TB infections (Glickman *et al.*, 2000; McKinney *et al.*, 2000; Fritz *et al.*, 2002; Movahedzadeh *et al.*, 2004; Munoz-Elias and McKinney, 2005; Miner *et al.*, 2009; Singh *et al.*, 2009). It is perhaps not surprising therefore that several modelling efforts in tuberculosis have focused on metabolism.

Much of what is known about metabolism in *M. tuberculosis* has been gleaned from conventional biochemical and molecular studies over many decades. The pathogen appears typical of bacteria of the Actinomycetales order, with a predominantly aerobic metabolism that is able to catabolize a wide range of substrates to generate biomass and energy. The genome encodes all the enzymes of the Embden–Meyerhof–Parnas pathway (EMP) and pentose phosphate pathway (PPP) and has a complete, or nearly complete TCA cycle (see below). The pathogen also encodes a functional glyoxylate shunt as well as several enzymes connecting the TCA cycle and glycolysis that may be used for either anaplerosis or gluconeogenesis.

There are, however, several features of central metabolism in *M. tuberculosis* that appear to be unusual. Although the link between glycolysis and the TCA cycle is complete in *M. tuberculosis,* the closely related pathogen *M. bovis* lacks a functional pyruvate kinase, and is therefore unable to deliver sugars from glycolysis to the TCA cycle. It is thus unable to utilize carbohydrates as the sole carbon source (Keating *et al.*, 2005). This function is therefore unnecessary *in vivo*, as this pathogen causes very similar disease in humans to *M. tuberculosis.* The role of isocitrate lyase has been intensively studied since the demonstration that both of the isocitrate lyase genes encoded by this pathogen, *icl*1 and *icl*2 (although some strains only have *icl*1) play an essential role in virulence (McKinney *et al.*, 2000; Munoz-Elias and McKinney, 2005). This finding has been generally interpreted to be due to this enzyme's role in the glyoxylate shunt and a metabolic shift in the principal carbon source from carbohydrates to fat in the host. However the role of the isocitrate lyases maybe more complex than just fat catabolism, as these enzymes also function as methyl citrate lyases in the methyl citrate cycle (Munoz-Elias *et al.*, 2006), which is used to catabolize propionate, derived from the oxidation of odd-numbered chain and branched chain fatty acids. In addition, it has also been demonstrated that the glyoxylate shunt functions concurrently with an oxidative TCA cycle that is completed by an anaerobic α-ketoglutarate ferredoxin oxoreductase (KOR) (Baughn *et al.*, 2009).

The TCA cycle also seems to be atypical since α-ketoglutarate dehydrogenase (KDH) activity has not be detected so the pathogen appears to lack the standard connection between α-ketoglutarate and succinyl CoA (Baughn *et al.*, 2009). Initially the cycle was proposed to be completed by an alternative route involving α-ketoglutarate decarboxylase (Kgd) to produce succinic semialdehyde, which is then converted to succinate by succinate semialdehyde dehydrogenase (Tian *et al.*, 2005). However, such a pathway left open the question of how succinyl CoA, an essential precursor for methyl-branched lipids, diaminopepimelate, sulfolipids and haem, is formed. Recently however, the gene proposed to encode KDG (Rv1248c) was actually found to encode a 2-hydroxy-3-oxoadipate synthase, which catalyses C–C bond formation between the activated aldehyde of alpha-ketoglutarate and the carbonyl of glyoxylate to yield 2-hydroxy-3-oxoadipate (de Carvalho *et al.*, 2010). The role of this activity in the metabolism of *M. tuberculosis* remains unclear. An alternative means of completing the TCA cycle has recently been uncovered with the discovery of anaerobic-type α-ketoglutarate ferredoxin oxidoreductase capable of interconverting between α-ketoglutarate and succinyl CoA, an enzyme activity that is often present in microaerophilic and strictly anaerobic organisms that run a reductive TCA cycle. However, as we will discover below, it is possible that the TCA cycle may not be complete in *M. tuberculosis*, at least under some circumstances. Fig. 7.1 illustrates the central metabolic pathways of *M. tuberculosis*, as understood in 2011.

Experimental systems for systems biology

Systems biology is an iterative procedure of experimental data acquisition, model building, hypothesis generation and experimental verification. One of the constraints upon this approach surrounds the experimental basis of this work. Models should be developed and validated with accurate and reproducible data. Moreover, the mathematical underpinning of many modelling approaches such as ^{13}C-MFA have an absolute requirement for the cultivation of the organism under steady-state conditions where metabolite concentrations are maintained at constant levels. This makes it very difficult to apply these approaches directly to pathogens such as *M. tuberculosis* growing *in vivo*, as such steady states are not attainable in mammalian cells. However, a standard approach in systems biology is to initially study systems in highly controlled experimental environments that allow models to be parameterized before their subsequent application in real life situations. One of the pioneers of systems

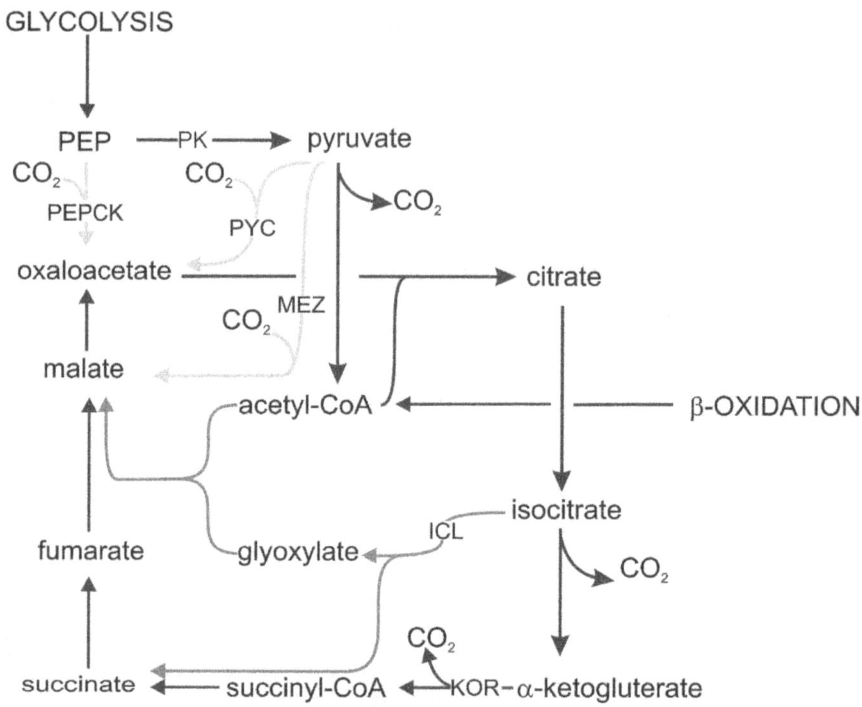

Figure 7.1 A metabolic map of central metabolism in *M. tuberculosis* showing the reactions surrounding the TCA cycle. The standard TCA cycle and the variant (SSA) pathway are shown. Anaplerotic/gluconeogenic reactions and the glyoxylate shunt are shown. Only enzymes mentioned in the text are indicated, including KOR (α-ketoglutarate ferredoxin oxireductase), KGD (α-ketoglutarate decarboxylase), GabD1/D2 (succinic semialdehyde dehydrogenase), ICL (isocitrate lyase) and MEZ [malic enzyme (malate dehydogenase, decarboxylating)], PEPCK (phosphoenolpyruvate carboxykinase), PK (pyruvate kinase) and PYC (pyruvate carboxylase).

biology, Hiroaki Kitano (Kitano, 2007), uses an example from racing car design to illustrate this approach. Cars are initially designed using a computer and then tested in a wind tunnel before being deployed in the actual race. By controlling airflow, wind tunnels transform a highly dynamic unsolvable system into steady-state that is amenable to mathematical modelling. Kitano argues that systems biologist's similarly needs biological 'wind tunnels' to develop their models. We here describe the application of one of the few biological wind tunnels: the chemostat.

Traditional batch cultivation remains the standard for most microbiological investigations. Typically, the microbe is inoculated in to a stirred vessel filled with rich media. The organism will grow at close to maximal rate (logarithmic phase) until either nutrients (including oxygen) become limiting or inhibitory products accumulate to levels that retard growth (stationary phase). Although convenient and suitable for many microbiological, genetic and functional genomic applications, this culture method is unsuitable for most systems biology applications, because (a) it is dynamic with cells adapting to a constantly changing environment; (b) it is not usually possible to monitor rates of substrate utilization or product accumulation; (c) the culture system is usually uncontrolled and thereby subject to wide fluctuations in parameters such as pH or oxygen concentration; (d) several micro-environments exist in most batch culture vessels that allow microbial growth in different physiological states so that the average value of measured parameters may not represent the actual value of those parameters in any single cell (the mode and the mean are very different so no cell exists with the actual parameter values obtained from measurement). This last consideration makes modelling of batch culture systems extremely problematic.

The need for maximal control of the experimental aspects of systems biology together with attainment of steady-state conditions has led to resurgence in the use of continuous culture systems such as chemostats (Hoskisson and Hobbs, 2005). During continuous culture in a chemostat microbes are grown at a rate set by the experimenter and other environmental parameters such as pH, oxygen levels are also precisely controlled. Culture medium is pumped at a constant rate into the vessel whilst the volume of the culture is kept constant by an overflow system. The flow rate (f) of the media is set by the experimenter to give a desired dilution rate (D). The dilution rate is the number of culture volumes passing through the chemostat per unit of time and equals the flow rate divided by the culture volume (V). The chemostat controls growth rate (μ) by limiting the availability of a growth substrate. The medium contains a fixed concentration of the limiting substrate, all the other nutrients being present in essentially excess amounts. By adjusting the feed rate the growth rate can be adjusted to 1–90% of the maximum growth rate for the organism. When a dilution rate is set the cells will initially grow as in a batch culture at the maximum specific growth rate (μ_{max}) until a substrate in the medium becomes limited. Eventually the cells adjust to the rate of nutrient supply so that the specific growth rate equals the dilution rate, i.e. $D = \mu$. This balanced growth is known as steady state and may be maintained indefinitely. At steady state the physiology of the cells remains constant, cellular processes being controlled by the concentration of the limiting substrate.

The chemostat therefore enables growth in a tightly regulated steady-state environment and thereby eliminates the inherent variability and dynamics of constantly changing batch cultures. The chemostat system is thereby analogous to the aerodynamic wind tunnel. It effectively freezes the dynamics of microbial growth to attain a steady-state system that is amenable to constraint-based modelling approaches, such as flux balance analysis, which critically depend on the assumption that concentrations of internal metabolites are held constant during the experiment. Data from chemostat cultivations are therefore more precise, reproducible and statistically significant than those obtained from batch cultivations (Hayes *et al.*, 2002; Boer *et al.*, 2003, 2007). Moreover, because the cultures are relatively homogenous, the mean value of measured parameters in samples removed from the chemostat is likely to correspond to the mode value of those parameters in the bulk population; allowing application of these values for model parameterization.

The slow growth rate of pathogenic mycobacteria, problems associated with clumping of bacilli and safety considerations have all provided obstacles for researchers attempting continuous cultures of *M. tuberculosis*. James and colleagues (James *et al.*, 2000) were the first to successfully cultivate *M. tuberculosis* in a chemostat at a growth rate of 24 hours in a complex nutrient-rich medium and have used their system successfully to investigate the responses of *M. tuberculosis* to oxygen (Bacon *et al.*, 2004) and iron limitation (Bacon *et al.*, 2007) and also mutation rates at different pH (Jenkins *et al.*, 2009). These studies demonstrated that the chemostat provides a reliable and reproducible environment for culturing mycobacteria and is therefore a very useful tool for 'omic' scale analysis such as DNA microarrays. It has been demonstrated that gene expression data from organisms, including *M. tuberculosis*, grown in the chemostat is significantly more reproducible than batch culture DNA-array data (Hayes *et al.*, 2002; Daran-Lapujade *et al.*, 2003; Bacon *et al.*, 2004).

Our group has developed a system for growing mycobacteria in a carbon-limited chemically defined minimal medium which can be used as a reproducible platform for systems biology studies (Beste *et al.*, 2005, 2007a,b, 2009). Initial studies using this experimental system to grow *Mycobacterium bovis* BCG (a non-pathogenic surrogate for *M. tuberculosis*) provided vital information on the biomass composition of the tubercle bacillus (Beste *et al.*, 2005). Studies prior to this are limited and were performed in poorly defined batch cultivations. For genome scale metabolic models the equations defining the biomass synthesis are very important and can impact on the predictive accuracy of the model. For two different growth rates ($D=0.03$, $t_d=23.1$, or $D=0.01$, $t_d=69.3$) the elemental and macromolecular composition of the biomass was measured, and shown to change as a function of the growth rate. This study demonstrated that more than half of the dry mass of the mycobacterial cell comprised carbohydrates and lipids, with only a quarter of the dry weight consisting of protein and RNA, but that these proportions change, depending on the growth rate. These data allowed a stoichiometric composition model for *M. bovis* BCG to be reconstructed, which is an important first step in the development of a metabolic network (Beste *et al.*, 2005).

Metabolic model building

The ultimate goal of system biology approaches to studying TB is to construct a complete model of infection incorporating both the pathogen and host, but this is currently infeasible as the information about the different components to be included in the model is lacking. Studies with other organisms have demonstrated that metabolism is, by far, the best understood cellular network and is thereby an excellent starting point for a systems-based approach (Bumann, 2009; Durot *et al.*, 2009; Feist *et al.*, 2009).

However, metabolism is complex. Even the simplest organisms synthesize many hundreds of metabolites connected by a similar number of enzyme-catalysed reactions. Each reaction is described by a set of kinetic parameters (e.g. K_m, V_{max}) which, in combination with substrate/product concentrations, determine its rate. Although K_m values are constants (for a particular substrate/product combination) and may be determined experimentally, intracellular concentrations of substrate, products and enzyme (influencing V_{max}) vary over wide ranges and are not easily measured. Even a single enzyme reaction is therefore a highly dynamic system; and systems of just a few reactions steps are usually mathematically described by a set of ordinary differential equation with a large number of parameters and variables whose values are extremely challenging to measure experimentally. Kinetic models have therefore only been applied to the dynamics of small well-defined systems, such as glycolysis in *Escherichia coli* (Bettenbrock *et al.*, 2006) that are very far from being genome-scale.

However, it is relatively straightforward to generate a metabolic network that describes the biochemical reactions that an organism is predicted to be capable of performing, in terms of stoichiometric formulas. The genome annotation predicts all those genes encoding enzymes that are readily linked to databases describing the predicted biochemical conversions performed by these enzymes. These stoichiometric reactions can then be incorporated into pathways that may be integrated into a network to construct a genome-scale metabolic network that effectively describes the chain of reactions responsible for inputting nutrients and transforming them into the biomass of the cell, waste products and energy.

On their own genome-scale metabolic networks are essentially descriptive, describing the set of metabolic reactions and their connectivity but cannot actually simulate any metabolic process. To do that they must incorporate metabolite concentrations and, for dynamic systems, all the kinetic parameters discussed above, which is infeasible at a genome scale. However, if a critical assumption is made, that all metabolite concentrations are held constant, then the underlying dynamics are hugely simplified and the system may be solvable. Consider the simple set of stoichiometric conversions, $A \rightarrow 2B \rightarrow C + D$. If this is a dynamic system then knowing the concentration of A tells us nothing about the concentrations of B, C or D without knowledge of the detailed kinetic parameters. Yet if the concentrations of A, B, C and D are unchanging (the system is at steady-state) then the rate of flux from $A \rightarrow 2B$ must equal the rate of flux from $2B \rightarrow C + D$ so knowing the concentration of A (or the flux towards A) uniquely determines the concentration (or the fluxes) of all the downstream metabolites. Steady state systems are thereby described by far fewer parameters and variables than dynamic systems.

Of course a genome-scale model is far more complex than the above system but the entire set of stoichiometric conversions for an organism is available from its genome annotation and standard biochemical literature. It is therefore possible to build a model consisting of all the stoichiometric reactions predicted by the annotation and link these pathways and networks connected by flux values between each reaction. Such a network can be incorporated into a mathematical model that once parameterized by measurements, such as substrate uptake or biomass production rates, can be solved by standard linear algebra tools (rather than the differential calculus required by kinetic models). However, because metabolic networks contain multiple branch points and parallel pathways there is not a unique solution but a vast space of possible solutions (the system is underdetermined). It is therefore necessary to apply constraint-based approaches, which reduce the solution space and thereby predict metabolic capabilities or internal fluxes (Duarte *et al.*, 2007; Feist *et al.*, 2007; Mo *et al.*, 2007; Oberhardt *et al.*, 2008; AbuOun *et al.*, 2009; Thiele *et al.*, 2009). Flux balance analysis (FBA) uses the procedure of optimization to reduce the solution space (for a review of FBA we recommend (Schilling and Palsson, 1998; Oberhardt *et al.*, 2009a; Orth *et al.*, 2010). FBA finds the flux solution that optimizes some parameter, which might be biomass production rate (and thereby growth rate), ATP synthesis, substrate consumption, product production or any other parameter of the model. Clearly there is a strong assumption in FBA that the cell applies a similar optimization strategy and thereby grows at its optimal growth rate, ATP production rate or rate of other optimized parameter. If that assumption is correct then FBA will find the correct solution – the one that the cell finds – and the FBA solution will correspond to the biological reality. It is of course an open question how often and in what circumstances microbial cells do optimize simplistic parameters such as growth rate, but FBA has been successfully applied for a variety of microbes in many situations but, most notably, in biotechnology applications where the aim of the research is indeed to optimize an output, such as product production.

Metabolic flux analysis (MFA) applies an alternative approach to reducing the solution space: applying additional measurements as constraints (Wiechert, 2001; Kim *et al.*, 2008). These may be measurements of intracellular metabolites, enzyme activities or indeed any additional measurement constraints, but the most powerful method currently available is

^{13}C-MFA, which derives solutions for the intracellular fluxes from the distribution of ^{13}C from a substrate into central metabolites and the amino acid products derived from central metabolites.

There are of course limitations to these approaches such as the requirement for steady or quasi-steady state conditions. Also, since no consideration is made of either transcriptional, translational, metabolic regulation or enzyme kinetics the predictive capabilities of constraint based models are limited to situations when these factors are not significantly influencing reaction rates (Durot *et al.*, 2009). Nevertheless, these approaches have been successfully applied to predict the metabolic capabilities of many different cellular systems (e.g. Varma and Palsson, 1994; Peng *et al.*, 2004; Kayser *et al.*, 2005; Antoniewicz *et al.*, 2007; Raman and Chandra, 2009; Boyle and Morgan, 2009). The application of both of FBA and MFA to *M. tuberculosis*, will be discussed below.

Metabolic models of *M. tuberculosis*

The first *M. tuberculosis* constraint based model was constructed by Raman *et al.* (2005) and consisted of all the reactions in mycolic acid synthesis (Raman *et al.*, 2005). This submodel of metabolism was composed of 219 reactions that involved 197 metabolites, catalysed by 28 enzymes. FBA was used to simulate mycolic acid metabolism and to identify potential drug targets in these pathways. The study illustrates the importance of optimization in FBA. As already discussed, FBA reduces the solution space by optimizing a parameter, usually known as the objective function, so choice of the choice of the parameter to be used as the objective function has considerable influence on the solutions obtained. Popular objective functions include: maximization or minimization of ATP production; maximization of redox potential; maximization of the rate of synthesis of a particular product, or minimization of nutrient uptake, but the most commonly used parameter is maximization of growth rate which has been successfully applied in many systems including nutrient limited chemostat culture of *E. coli* (Schuetz *et al.*, 2007). Its use is more problematic for slow growing pathogens, such as *M. tuberculosis*, since it has not been established that these organisms do actually maximize their growth rate. The study used two objective functions that optimized the production of mycolic acids. The first, termed C1, optimized production of only the most abundant mycolate; whereas the objective function C2 maintained the known ratios of different mycolates. To test the predictive accuracy of these objective functions *in silico* deletions were performed and compared with transposon site hybridization mutagenesis data. The highest predictive accuracy was obtained with the objective function C2 with an 82% correlation with experimental data. FBA identified 16 essential genes in this study and this primary list was then filtered to remove any genes encoding proteins that were complemented by homologues, and also those with close homologues in the human proteome. This feasibility analysis identified seven potential drugs targets for anti-TB drug design (discussed below).

Although targeting a small subsystem such as mycolic acid synthesis can yield valuable information on specific pathways, it has limited value in elucidating the metabolic capability of *M. tuberculosis*. This latter objective is best approached by constructing a genome-scale network of metabolism (Beste *et al.*, 2007a; Jamshidi and Palsson, 2007). The first published genome scale network was built by our group using *Streptomyces coelicolor* as a starting model (Beste *et al.*, 2007a). Orthology relationships were mapped between the related species using the KEGG databases and this preliminary model was further supplemented with data from the BioCyc database. This automatic process, however accounted for only 57% of the final model. The remaining model was reconstructed by labour intensive manual curation based upon primary research publications, textbooks, and review articles, and also by picking the brains of experts in the field. The final model utilized two biomass formulations which were derived from published data of cell composition obtained from a variety of sources, including our own chemostat-derived data obtained from fast- and slow-growing BCG. BIOMASS 1 reflects the actual macromolecular composition of *in vitro*-grown *M. tuberculosis*; whereas BIOMASSe consisted of only those cellular components, such as DNA, RNA, protein, co-factors and the cell wall skeleton, that were considered to be essential for *in vitro* growth. The advantage of having these two biomass formulations is that the model could be used to predict gene essentiality both *in vitro* (with the minimal BIOMASSe as the objective function) and *in vivo* (with the more complete BIOMASS 1 including many virulence factors as the objective function).

The final functional genome scale metabolic network of *M. tuberculosis* (GSMN-TB) consisted of 739 metabolites participating in 849 reactions and involves 726 genes. The model is freely available as both an Excel file or in sbml format; and is accessible via a user-friendly web tool for constraint-based simulations (http://sysbio3.fhms.surrey.ac.uk/). FBA-based predictions of *in vitro* gene essentiality using BIOMASSe as the objective function correlated well with predictions of gene essentiality obtained by global transposon mutagenesis (Sassetti *et al.*, 2003), with an overall predictive accuracy of 78% (Beste *et al.*, 2007a).

Quantitative validation of the model was also performed using data from continuous culture chemostat experiments (Sassetti *et al.*, 2003). The model predicted a lower rate of glycerol consumption than the experimentally determined values. A plausible explanation for the discrepancy was that, in addition to consumption of glycerol, the tubercle bacillus also utilized oleic acid released from hydrolysis of the Tween 80 dispersal agent present in the media. Opening an additional oleic acid transport flux corrected this discrepancy and we have recently confirmed that Tween 80 is indeed being consumed in these experiments.

A second genome scale reconstruction of *M. tuberculosis* was published Jamshidi and Palsson (2007). Silicon *iNj661* weighs in at 939 reactions and 661 genes and was constructed using several databases combined with manual curation (Jamshidi and Palsson, 2007). Comparing *in silico* predictions of growth rate and gene essentiality, with previously published experimental data, validated this model. The *in silico* predictions of the growth rate of *M. tuberculosis* on three different media agreed with experimental data but the model had a poor predictive accuracy for experimental essentiality data (55%). Interestingly the model predicted that *M. tuberculosis* is unable to grow with glucose as a sole carbon source, which has been observed for some strains of *M. tuberculosis*. It should however be emphasized that many studies into the growth of *M. tuberculosis* on 'sole carbon sources' are difficult to interpret as experiments are they have often been conducted in minimal media containing in addition to the 'sole carbon source' asparagine or glutamate as a nitrogen source and Tween 80 as a dispersal agent. There is experimental evidence to show that *M. tuberculosis* can utilize both Tween 80 (Wayne, 1994) and also asparagine (Lyon *et al.*, 1969) as a carbon source, so these experimental results should be treated with caution.

Building upon the work described by Ramen *et al.* (2005) hard-coupled reaction (HCR) sets were calculated for the iNj66I model in order to predict novel drug targets on a genome scale (Jamshidi and Palsson, 2007). Hard coupling reaction (HCR) sets are generated by a method which does not require an objective function and therefore overcomes any bias introduced by using a specific objective function to constrain the solution space. HCRs are groups of reactions that, due to mass conservation and connectivity constraints, must operate in unison. This strategy identified known and also potential new drug targets that require further analysis.

The mycolic acid synthesis submodel and the two genome scale network reconstructions available for *M. tuberculosis* illustrate the different approaches which can be applied to reconstructing, validating and applying metabolic models. They also provide a reference for future metabolic reconstructions. The next challenge is to combine these three models and build upon them by integrating any new experimental data in order to expand and refine the reconstructions in an iterative cycle. In this way the model can serve as an up-to-date representation of the cumulative knowledge of *M. tuberculosis's* metabolic capabilities. For comparison, the *E. coli* genome scale network has undergone six different successive reconstructions over the last 18 years, each one contributing positively to a large number of different studies (Feist and Palsson, 2008). A well-curated reconstruction is a prerequisite for all systems biology approaches to studying *M. tuberculosis*.

Metabolic models of host–pathogen systems

M. tuberculosis is an intracellular pathogen that replicates primarily in the phagosome compartment of macrophages so its biology is intimately connected to that of its host cell. To simulate the combined system, Bordbar and colleagues (2010) built a novel metabolic model that integrated the iNj66I model of *M. tuberculosis* with a cell-specific alveolar macrophage model, iAB-AMØ-1410 (based on the global human metabolic reconstruction, Recon 1; Duarte *et al.*, 2007) to build an integrated host–pathogen genome-scale reconstruction, iAB-AMØ-1410-Mt-661. The combined model was essentially composed of three compartments representing the macrophage, the phagosome and the pathogen residing within the phagosome. These were connected via metabolite and gas exchange reactions that allowed the *M. tuberculosis* compartment to uptake substrates and excrete waste products into the phagosome compartment (Fig. 7.2). The exchanges do of course instantiate several assumptions regarding the infectious state. The macrophage was assumed to be consuming glucose, glutamine and essential amino acids and excreting lactate. The phagosome environment that provided resources for *M. tuberculosis* replication was assumed to be depleted in glucose and rich in glycerol and fatty acids. A key aspect of the reconstruction was the biomass composition of both macrophage and *M. tuberculosis* compartments, remembering of course that biomass composition plays a key role in FBA and is often used as the objective function and thereby has a very substantial influence on the flux solutions. Macrophages do not readily multiply so the iAB-AMØ-1410 biomass reflected only maintenance function, such as lipid, protein, mRNA turnover, DNA repair, and ATP maintenance. With this objective function the macrophage model successfully predicted experimentally observed rates of glucose oxidation and lactate production. To provide the biomass equation for intracellular *M. tuberculosis* the authors examined gene expression data derived from *in vivo* mouse model studies as well as *in vitro* studies that aimed to mimic the infection environment. They then adjusted the *M. tuberculosis* biomass composition to optimize the fit to the gene expression data. This involved increasing the amount of amino acids, mycolic acids, mycobactin, mycocerosates and sugars in the biomass equation, reducing ATP maintenance, DNA, fatty acids

A

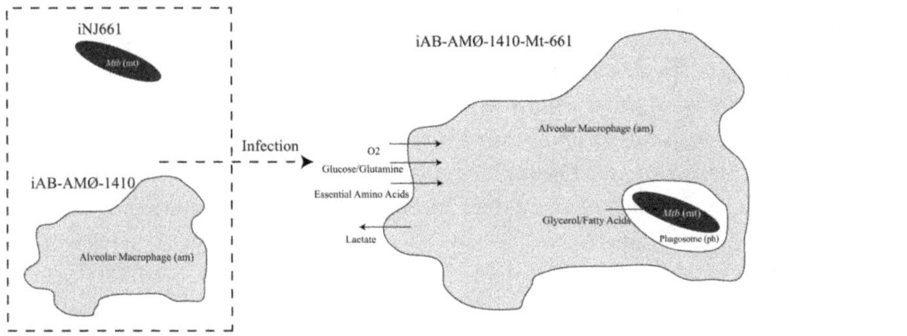

B

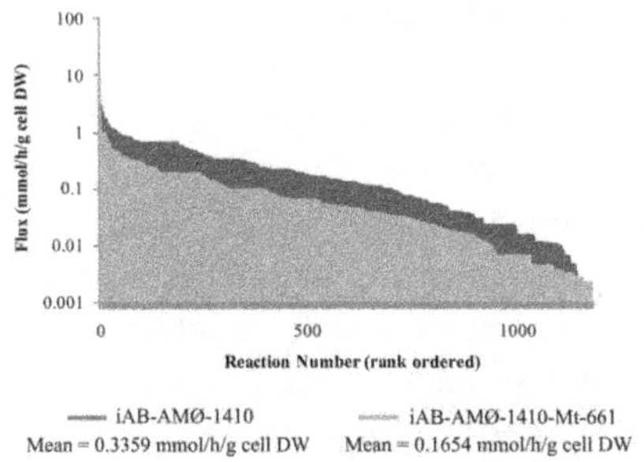

C

	iAB-AMØ-1410	iNJ661	iAB-AMØ-1410-Mt-661
Genes	1410	661	2071
Reactions (intracellular)	3012	939	3951
Reactions (exchange)	382	88	538
Gene Associated Reactions	69.3%	77%	71.1%
Metabolites	2572	828	3400
AMØ Max Biomass	0.0270	-	0.0238
AMØ Max ATP Production	0.6843	-	0.1703
AMØ Max NO Production	0.0359	-	0.0163
AMØ Max NADH Production	1.7806	-	0.5054
Original Mtb Max Biomass	-	0.0522	0.0021

Figure 7.2 Results obtained by integration of the alveolar macrophage (iAB-AMØ-1410) and *M. tuberculosis* (iNJ661) reconstructions. (A) Metabolic links between the extracellular space (e), alveolar macrophage (am), phagosome (ph), and *M. tuberculosis* (Mtb) in iAB-AMØ-1410-Mt-661. The model is compartmentalized using the abbreviations as shown. In the model, the major carbon sources of the alveolar macrophage were glucose and glutamine. The macrophage compartment was also aerobic and requires the essential amino acids. Despite its use of oxygen, the macrophage exhibits anaerobic respiration and produces excess lactate. In the *M. tuberculosis* compartment of the model, the major carbon sources available in the phagosome environment were glycerol and fatty acids. The phagosome environment was also functionally hypoxic. (B) The flux span of iAB-AMØ-1410-Mt-661 was significantly reduced (51%) compared with its progenitor macrophage model, iAB-AMØ-1410. This shows a stricter definition of the alveolar macrophage solution space without adding additional constraints on the alveolar macrophage portion of the network. (C) Reaction, metabolite, and gene properties of the three reconstructions. Maximum production rates of ATP, nitric oxide, redox potential (NADH), and biomass are shown. Adapted from Bordbar *et al.* (2010).

and phospholipids and removing peptidoglycans and phenolic glycolipids entirely from the biomass equation to construct a new objective function. It should be emphasized that the resulting behaviour of the reconstituted model is dependent on the precise composition of this adjusted biomass.

The authors then examined changes in flux state of the *M. tuberculosis* compartment as a consequence of its simulated replication in the macrophage (compared with the iNj66I model). The simulation predicted a shift in carbon uptake with suppression of glycolysis and up-regulation of gluconeogenesis, together with production of acetyl-CoA coming from macrophage-derived fatty acids via the glyoxylate shunt. Concomitant with the utilization of fatty acids as carbon source was up-regulation of fatty acid oxidation pathways. There was a shift towards mycobactin and mycolic acid synthesis with reduced flux through nucleotide, peptidoglycan, and phenolic glycolipid pathways. Many of these changes are likely to be a consequence of the altered biomass composition.

The accuracy of the model was tested by comparison of gene essentiality predictions of the *M. tuberculosis* component of the model with genes identified to be conditionally essential for infection in a mouse lung model (but not essential *in vitro*) by Transposon Site Hybridization (TraSH) (Sassetti and Rubin, 2003). 374 genes investigated by TraSH were in the model. Of these, the *in silico* analysis predicted that only 9 genes were conditionally essential in the *M. tuberculosis* compartment of the integrated model. Of those 9 *in silico*-predicted essential genes, only two genes were also essential experimentally by TraSH. Many of the discrepant results are likely to be due to differences between the simulated macrophage system and the mouse model that was used to obtain the TraSH data. Four of the nine genes were components of nitrate reductase, which has been shown to play a role in *in vitro* models of infection (Sohaskey and Wayne, 2003) and was thereby incorporated into the model when the gene expression data was used in the fitting exercise. However, even discrepant model predictions can be informative. Systems biology models are essentially a mathematical instantiation of biological hypotheses. In this case one of the hypotheses embedded in the iAB-AMØ-1410-Mt-661 model was that nitrate reductase was required for survival of *M. tuberculosis* in the mouse. This hypothesis was tested by comparison of model predictions with TraSH data (Sassetti and Rubin, 2003) and shown to be discrepant (Bordbar *et al.*, 2010). However, the situation is of course more complicated as nitrate reductase, although apparently non-essential in the mouse lung model (Sassetti and Rubin, 2003), has been shown to contribute to the virulence of BCG in immunodeficient mice (Weber *et al.*, 2000). It seems likely therefore that different models will have to be constructed for different host: pathogen combinations. Testing *in silico* predictions experimentally is therefore an essential step in the refinement and improvement of systems biology models in an iterative cycle of model → prediction → experimental test → model.

Applications of the models

Using models to interrogate genome annotation

Genome-scale networks are usually constructed initially from genome annotation and are thereby subject to errors in that annotation. However, the metabolic model scrutinizes the metabolic component of genome annotation at a system level for functionality and can thereby be used to find pathway holes or inconsistencies in the annotation. There are several 'orphan reactions' in GSMN-TB, that is reactions that are required for network functionality but for which there is no annotated *M. tuberculosis* gene predicted to perform that function. For example, sulfolipid synthesis in *M. tuberculosis* generates the metabolite adenosine 3′,5′-bisphosphate (PAP in the model) which will accumulate and thereby become toxic (unbalanced in the model) if it is not catabolized. The model is therefore infeasible unless the reaction catalysed by the enzyme 3′,5′-bisphosphate nucleotidase (which converts the metabolite to AMP and inorganic phosphate) is included in the network, as an orphan reaction. Examining model feasibility thereby generates clues to incomplete or incorrect genome annotation and may even provide novel drug targets that are not apparent in the genome annotation.

In silico models also allow genome annotation to be scrutinized by systems-based experimental data. For example, the route for glycerol utilization is generally assumed to proceed via glycerol kinase (encoded by *glpK*) followed by dehydrogenation; however, the genome annotation of *M. tuberculosis* includes several alcohol dehydrogenases that could be involved in an alternative uptake pathway whereby glycerol is first oxidized by glycerol dehydrogenase before being phosphorylated (this pathway is annotated in the KEGG *M. tuberculosis* pathway map). However, incorporation of this pathway into the initial GSMN-TB model led to the prediction that the gene *glpK* is dispensable for growth on media with glycerol as sole carbon source. Global mutagenesis data demonstrated that *glpK* was in fact essential for growth on glycerol, which was confirmed by construction of a single *glpK* knock-out mutant (Beste *et al.*, 2009). This information was incorporated into a refined GSMN-TB model in which the annotated alcohol dehydrogenases do not provide an alternative glycerol uptake pathway.

Other systems-based insights into the metabolism of *M. tuberculosis* can be obtained by simply performing growth simulations with the model. For instance, it is often claimed that *M. tuberculosis* requires operation of the glyoxylate shunt for growth on lipids. However, FBA-based simulation of GSMN-TB indicated that although the isocitrate lyase reaction of the glyoxylate shunt is predicted to be essential for growth on simple fatty acids such as acetate, it was not predicted to be essential for growth on complex lipids, such as phospholipids. The reason is apparent on examination of the flux solution: catabolism of phospholipids yields glycerol as well as acetate, which can be used for gluconeogenesis without operation of the shunt.

Drug discovery

Both the GSMN-TB model and iNj661 were successful in predicting the known targets of current antituberculosis drugs as essential genes. But although gene essentiality is a likely prerequisite for a drug target it is not sufficient. Many other factors need to be taken into account such as the presence of human homologues or existence of inhibitors: the *drugability* of the target. Raman and colleagues (Raman *et al.*, 2008) developed TargetTB as a platform for drug discovery. The system incorporates all three published *M. tuberculosis* metabolic models to impose sequential filters on an initial list of essential genes in order to generate several shortlists of putative drug targets. The first list was generated by combining essentiality data, from a protein–protein interactome and from FBA studies, using both the available genome scale models and also the mycolic acid pathway (MAP) submodel. This list was then filtered using sequence analysis to remove genes with homologues in humans; structural analysis to remove targets with similar binding sites to human proteins and this process identified 622 putative drug targets. Further filtering generated a short list of broad spectrum antibacterial, TB specific targets and also those candidates that could be important for treating persistent TB (216 targets). Several known and proposed candidate drug targets were identified using TargetTB, validating this approach and indicating that the other targets on this list could also be promising candidates for subsequent experimental investigations.

Kim and colleagues (2009) used an alternative metabolite-centric approach to drug target identification for a number of pathogens including *M. tuberculosis*. Their strategy was to use metabolite essentiality and *chokepoint analysis* (Kim *et al.*, 2010), which identifies enzymes that uniquely produce and/or consume a particular metabolite. Metabolites that were both essential and chokepoints were then filtered to remove metabolites with human homologues. This analysis identified 413 essential metabolites, 554 chokepoints and a final list of 364 putative drug targets in *M. tuberculosis*. Mycolate biosynthesis was identified as a potential target, which is of course targeted by current anti-tubercular drugs. Another potential metabolite was identified as chorismate, a product of the shikimate pathway which is essential in higher plants, fungi, bacteria and algae but absent in mammals. Chorismate biosynthesis has been independently identified as an attractive drug target in several systems (Dias *et al.*, 2007; Ely *et al.*, 2008).

Yet another approach was developed by Raman and colleagues. (2009). Their study (Raman *et al.*, 2009) identified protein–protein dependencies in the *iNj661* network to calculate a *metabolic disruptability index* for each protein in the network. This allowed them to identify hub proteins that play a dominant role in the network such that inhibition of their action would achieve maximal metabolic disruption. Confirmation of the metabolic impact of hub protein disruption was obtained by FBA, which demonstrated that most of the identified hub proteins (and combinations of proteins) are also synthetic lethals *in silico* and may therefore be good drug targets.

A different strategy was employed by Fang and colleagues, who combined FBA-based growth simulations of *iNj661* with kinetic modelling of enzyme action and cell growth dynamics (Fang *et al.*, 2009). Two metabolic inhibitors, 3-nitropropionate and 5′-O-(N-salicylsulfamoyl) adenosine, which target the enzyme isocitrate lyase and salicyl-AMP ligase respectively, were chosen to test the model. These enzymes had already been identified as potential targets for therapeutic intervention and experimental data. Isocitrate lyase has been intensively investigated as a drug target by virtue of its *in vivo* role in both the glyoxylate shunt and methyl citrate cycle (Munoz-Elias and McKinney, 2005) and the methylcitrate cycle (Munoz-Elias *et al.*, 2006). Salicyl-AMP ligase plays a vital role in the synthesis of the mycobactin siderophore that is essential to intracellular survival (De Voss *et al.*, 2000). Kinetic data on the action of the inhibitors was also available. The analysis demonstrated that the model successfully predicted the experimental dose response curve of both these drugs.

In summary, both genome-scale metabolic models and smaller scale kinetic models have been used for drug target discovery and simulation of drug action in *M. tuberculosis*. However, it is also fair to say that systems biology approaches have yet to deliver new validated drug targets for this pathogen (or indeed any pathogen). However, it may be that the biggest impact of systems biology will be further down the drug discovery pipeline. Most experimental drugs never make it to market because of unwanted side-effects discovered only after extensive and expensive trials in both animal model and in the clinic. If systems biology could be used to predict side-effects then it could become an enormously powerful tool for drug development. Clearly this would require modelling of both the pathogen and the host. However, as described above, progress has already been made in this area with the construction of a host–pathogen model for the replication of *M. tuberculosis* in macrophages.

Interpretation of experimental data

The functional genomics revolution has provided the means to generate high-throughput datasets but integration and interpretation of vast numbers of data points to generate new hypotheses remains a formidable challenge. Computational models can serve to bridge the gap between data and hypothesis driven research providing a framework for integration of high throughput data that can lead to model revisions (to resolve discrepancies between model predictions and experiment) and semi-automatic generation of new hypotheses (Covert *et al.*, 2004; Oberhardt *et al.*, 2009b). Even simply overlaying 'omic' data onto genome scale metabolic models provides a metabolic context to interpret this data and can also highlight incomplete or incorrect knowledge (Oberhardt *et al.*, 2009b).

Gene essentiality data

One of the most straightforward applications of genome-scale models is to predict essential genes that can then be compared with experimental data. For example, using a high-throughput transposon site hybridization screen we identified the genes essential for the growth of *M. bovis* BCG on glycerol and compared this with gene essentiality predictions using GSMN-TB (Beste *et al.*, 2009). Whilst there was a good correlation between the GSMN-TB and the experimentally observed gene essentialities (76.66%) the analysis demonstrated how the model could be used to highlight gaps in our knowledge of mycobacterial metabolism. Some of the discrepancies can be attributed to an undefined level of inaccuracy in global mutagenesis assays but may also be due to gene regulation of isoenzymes. For instance, both menaquinol oxidase systems (the aa3-type and bd-type) are predicted to be non-essential as they are functionally redundant in the model. This contradicts the global mutagenesis data, which indicated that the aa3-type cytochrome c oxidase is in fact essential, and likely to keep the main electron transport system operating under aerobic conditions.

Transcriptome data

Whereas it is relatively straightforward to obtain multiple measurements to define the physiological or metabolic state of bacteria *in vitro*, only limited information can be obtained for bacteria *in vivo*. In particular, it is very challenging to perform metabolomic, proteomic, biochemical, physiological or structural studies with the small numbers of organisms obtained from infected animals. However, it is possible to perform transcriptomic studies on *in or* ex *vivo* grown organisms and these methods have been applied to the TB bacillus to obtain transcriptomic profiles of bacteria growing in cultured macrophages, mouse models and in human lesions (Schnappinger *et al.*, 2003; Talaat *et al.*, 2004; Cappelli *et al.*, 2006; Raju *et al.*, 2008; Tailleux *et al.*, 2008). The transcriptional profile of a cell (via translation, enzyme activity etc.) determines most aspects of its physiological state; therefore it should be possible to predict a physiological state from knowledge of the complete cellular transcriptome. However, the mapping between messenger RNA levels and physiological state is highly complex and non-linear depending on many unknown factors such as mRNA stability, translation efficiency and post-translational modification of proteins. Traditional approaches to defining metabolic responses from transcriptome data have generally relied on examining expression levels of key (rate-controlling) genes in metabolic pathways (for instance, Salmon *et al.*, 2003). However, metabolic control analysis has demonstrated that control is distributed throughout the entire metabolic network, such that the flux through any particular pathway is controlled globally (Heinrich and Rapoport, 1974; Kacser and Burns, 1995) rather than by a particular enzymatic step. This makes a simple mapping of differentially expressed genes onto metabolic pathways an unrealistic strategy for successful predictions of global metabolic state changes.

Several system-level approaches have been proposed to extract metabolic information from gene expression profiles. In the reporter metabolites approach (Patil and Nielsen,

2005) the local connectivity of a metabolite in the bi-partite, substance/reaction graph is used to identify a set of genes associated with each metabolite. Subsequently, for each of the metabolites, the distribution of the microarray-derived signal of genes associated with the metabolite is compared with the background distribution of the microarray-derived signal for all genes, resulting in the identification of the transcription regulation focal points of metabolism: network nodes that are directly affected by clusters of differentially expressed genes. In another approach, Shlomi *et al.* (2008) used mixed integer linear programming to minimize the discrepancy between the internal metabolic flux distribution and the transcriptional profile of genes encoding metabolic enzymes. Their approach identifies flux distributions, which are consistent with the stoichiometric constraints of the genome scale metabolic reaction network and at the same time maximize the number of active metabolic fluxes associated with up-regulated genes and the number of non-active metabolic fluxes associated with down-regulated genes. Yet another approach, E-flux, was recently developed and used to examine *M. tuberculosis* microarray data in the context of both the genome scale metabolic reaction network, by constraining upper bounds of metabolic reactions to values proportional to the microarray signals of genes associated with these reactions (Colijn *et al.*, 2009). Two models were used for the analysis: the Raman *et al.* (2005) model of mycolic acid pathways (Raman *et al.*, 2005) and the GSMN-TB genome-scale metabolic model (Beste *et al.*, 2007a). E-flux was applied to microarray data obtained from a large study that investigated the response of *M. tuberculosis* to 75 different drugs, drug combinations, and nutrient conditions (Boshoff *et al.*, 2004). Eight of the tested drugs target mycolic synthesis, and this was correctly predicted by E-flux analysis of the microarray data, indicating that the method may be useful for target identification of novel inhibitors.

Shi and colleagues (Shi *et al.*, 2010) applied a similar *in silico* method to interrogate quantitative PCR (qPCR) transcriptome data obtained from a model of respiratory infection of mice in which *M. tuberculosis* replicates in the lung for approximately 20 days, followed by stabilization of bacterial numbers due to expression of acquired cell-mediated immunity. The qPCR data were not genome-scale but focused on a set of genes predominantly involved in the pathways of central metabolism and lipid synthesis. These data were first interpreted qualitatively. Observed changes in mRNA abundance suggested that as tubercle bacilli stop replicating in the mouse lung and respond to the decreased demand for energy and biosynthetic precursors by down regulating glycolysis, PPP and the TCA cycle. The main function of central metabolism appears to shift from providing energy and biosynthetic precursors for bacterial growth to accumulating the storage compounds such as triacyl glycerides (TAG) and glutamate. To gain a genome-scale insight into the underlying metabolic changes, two *in silico* cells were constructed by adjusting biomass composition of the GSMN-TB model. One cell represented growing *M. tuberculosis*, while the other represented the more minimal cell composition predicted for non-growing *M. tuberculosis* in the mouse lung. FBA was then used to predict the consequent changes in flux distribution in the cell. The resulting flux distributions were broadly consistent with the gene expression data and the hypothesis that growth arrest in the mouse lung is associated with a re-outing of carbon flow in central metabolism from metabolic pathways generating energy and biosynthetic precursors to pathways for storage compounds, such as TAG and glutamate.

Our own laboratory developed an alternative method, differential producibility analysis (DPA). The method (Bonde *et al.*, 2011) relies on FBA to link genes with metabolites on a system-wide level. A gene essentiality scan is first performed on every gene but instead of using biomass as the objective function, each metabolite is, in turn, used as the objective function. It is thereby possible to generate a map that identifies, for each metabolite, the genes that are required for its synthesis (the *producibility plot*). In the next step of DPA, the experimental data are interrogated. Gene expression signals for a particular experiment are assigned onto each gene which, using the producibility plot, are mapped onto each metabolite. Metabolites are then ranked to identify the metabolites that are *most affected* by genes that are upregulated in the target experiment and (separately) are ranked to metabolites that are *most affected* by genes that are downregulated in the same experiment. The whole procedure of DPA effectively transforms a gene-based transcriptome signal into a metabolite-based metabolome signal.

The DPA method was first tested with *E. coli* transcriptome data and shown to successfully identify metabolic responses to environmental perturbation (shift to anaerobic growth) and gene knock-out. This method was then applied to several *M. tuberculosis in vitro* transcriptome datasets and was able to identify metabolic responses. Applying DPA to transcriptomic data obtained from *M. tuberculosis* replicating in mice derived macrophage (Schnappinger *et al.*, 2003) revealed a previously unrecognized feature of the response of *M. tuberculosis* to the macrophage environment (Bonde *et al.*, 2011): a down-regulation of genes influencing metabolites in central metabolism and concomitant up-regulation of genes that influence synthesis of cell wall components and virulence factors (Fig. 7.3). DPA suggests that a significant feature of the response of the tubercle bacillus to the intracellular environment is a channelling of resources towards remodelling of its cell envelope, possibly in preparation for attack by host defences. Interestingly, application of DPA (Bonde *et al.*, 2011) to transcriptome data obtained from *M. tuberculosis*

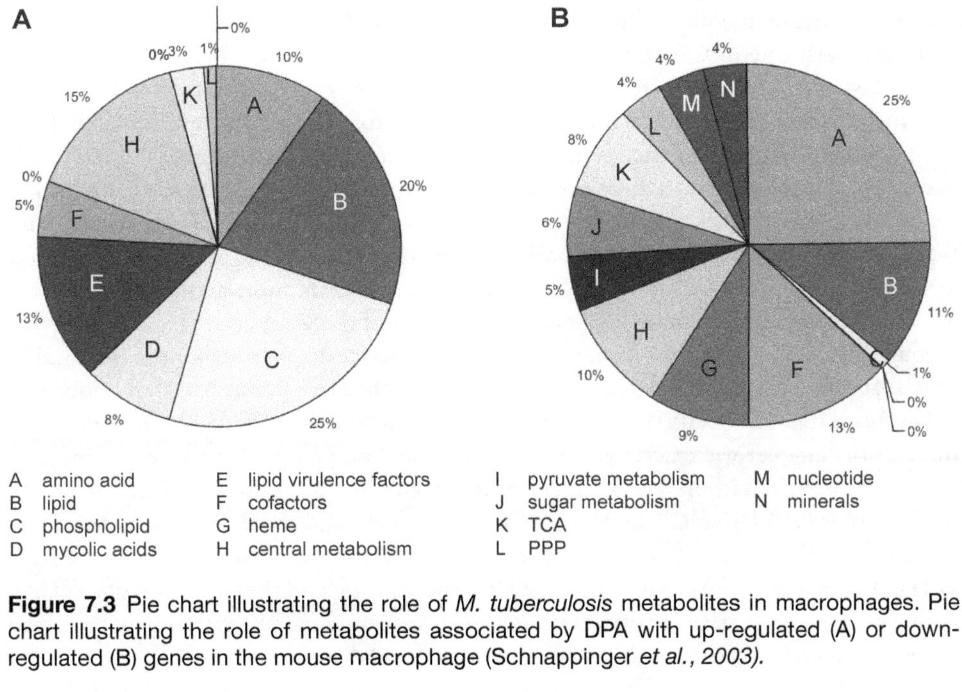

Figure 7.3 Pie chart illustrating the role of *M. tuberculosis* metabolites in macrophages. Pie chart illustrating the role of metabolites associated by DPA with up-regulated (A) or down-regulated (B) genes in the mouse macrophage (Schnappinger *et al.*, 2003).

bacilli recovered from human sputum (Garton *et al.*, 2008) generated a very different metabolic signature to the mouse macrophage data. DPA may thereby be used to unravel the mechanisms of virulence and persistence of *M. tuberculosis* and other pathogens and may have general application for extracting metabolic signals from other '-omics' data.

^{13}C-metabolic flux analysis (^{13}C-MFA)

The goal of metabolic analysis is to understand the metabolic pathways that are being utilized under particular conditions. As described above, it is possible to obtain estimates of the ranges of fluxes that are compatible with substrate inputs and outputs in a system using FBA. However, there are usually a great number of possible flux solutions that are compatible with the data so FBA utilizes the method of optimization to determine the flux distribution that optimizes some parameters, such as growth rate. This method has been very successfully applied in fast-growing organisms but its application to the slow-growing pathogen, *M. tuberculosis* is uncertain. Our earlier work used FBA to predict flux distribution in fast and slow-growing *M. tuberculosis* with glycerol as sole carbon source and predicted an increased flux through the glyoxylate shunt at slow growth rate (Beste *et al.*, 2007a). This was surprising as the shunt is usually considered to be used solely for growth on two carbon compounds such as acetate and the essentiality of this enzyme for *in vivo* growth has been widely interpreted to indicate that the pathogen consumes host lipids (yielding acetate on beta oxidation) during growth inside the macrophage (McKinney *et al.*, 2000; Munoz-Elias and McKinney, 2005, 2006). We confirmed that activity of the enzyme isocitrate lyase (ICL, the key enzyme of the glyoxylate shunt) was indeed induced during slow growth rate but how and why the organism was utilizing the glyoxylate shunt during slow growth on glycerol was a mystery. The mystery was compounded when we constructed an ICL mutant of *M. tuberculosis* and demonstrated that the mutant was unable to grow at slow growth rate in a glycerol-limited chemostat (Beste *et al.*, 2011). To discover the metabolic pathways involved in slow (and fast) growth we recently, for the first time, performed ^{13}C metabolic flux analysis (MFA) on *M. bovis* BCG and *M. tuberculosis* at fast and slow growth rates. Tracer experiments were performed with steady state chemostat cultures supplied with ^{13}C labelled glycerol. Through measurements of the ^{13}C isotopomer labelling patterns in protein-derived amino acids and enzymatic activity assays we identified the activity of a novel pathway (termed the GAS pathway), that is used for pyruvate dissimulation in *M. tuberculosis*. This pathway is characterized by significant flux through the glyoxylate shunt and also through the carbon fixing anaplerotic reactions at the phosphoenolypyruvate (PEP)–pyruvate–oxaloacetate node combined with very low flux through the succinate–oxaloacetate segment of the TCA cycle (Fig. 7.4). The flux through the GAS pathway is increased at slow growth rate accounting for the essentiality of ICL at slow growth rate. An interesting feature of the GAS pathway is that it included a significant fraction of flux (far more than required for anaplerosis) going through one or more of the anaplerotic reactions between phosphoenolpyruvate/pyruvate and malate/oxaloacetate in the carbon-fixing direction. This prediction of ^{13}C-MFA was confirmed by feeding *M. tuberculosis* ^{13}C-labelled sodium bicarbonate and confirming that the pathogen is indeed able to incorporate this unusual carbon source into amino acids (Beste *et al.*, 2011). As the human host is abundant in CO_2 this finding and the operation of the GAS pathway, requires further investigation *in vivo* as carbon dioxide fixation may provide a point of vulnerability that could be targeted with novel drugs.

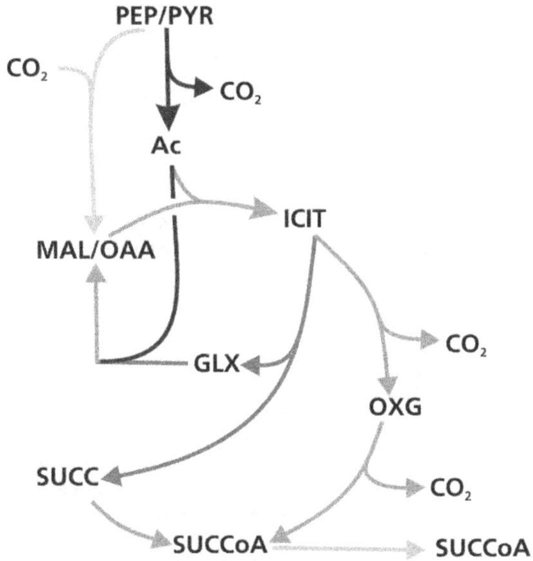

Figure 7.4 Schematic of the GAS pathway which is characterized by flux through the glyoxylate shunt and anaplerotic reactions for oxidation of pyruvate and succinyl CoA synthetase for the generation of succinyl CoA. Metabolite abbreviations: PEP/PYR phosphoenolpyruvate/pyruvate, Ac, acetate; CHO, ICIT, isocitrate/citrate; MALOAA, L-malate-oxaloacetate; OXG, 2-oxoglutarate; SUCC, succinate; SUCCoA, succinyl-CoA, GLX, glyoxylate.

Future challenges

The application of systems biology to the study of tuberculosis is a science that is still in its infancy. Nevertheless, significant progress has already been made. Several *in silico* models of *M. tuberculosis* have been constructed and a reconstruction of the *M. tuberculosis*-macrophage system has been described. The model building process itself is a highly informative exercise that not only defines a minimal metabolic capacity necessary for making a *M. tuberculosis* cell but also provides clues to gene annotation and generates novel insights into the metabolic capability of this pathogen. The models have been shown to be useful tools for drug target prediction. One of the most powerful applications of these approaches has been to use the *in silico* models to interrogate experimental data to provide system-level insight into underlying metabolic responses associated with the response of *M. tuberculosis* to stress, drugs and growth in host cells. Each of these models is available online giving researchers across the world access to systems biology tools that can be used to investigate the biology of the tubercle bacillus but also to interrogate both published and new datasets.

There is no question that existing models are not yet realistic reconstructions of the *M. tuberculosis* cell. However, interrogating the models with experimental data in an iterative cycle of model refinements and experimentation will ensure that the current models will become more accurate descriptions of the metabolism of *M. tuberculosis*. A limitation of current FBA-type models is that they can only strictly be applied to steady-state systems. A future goal will be to extend these models to simulate dynamic states, such as during *in vivo* growth. This will require integration of metabolic networks with gene regulatory networks and kinetic models of enzyme action together with realistic models of host–pathogen interactions. Such multiscale models may eventually be used to build an *in silico M. tuberculosis* cell. Such a model may be used for drug discovery and optimization of treatment regimes but will also be an enormously powerful tool to investigate the fundamental biology of this important pathogen.

References

AbuOun, M., Suthers, P.F., Jones, G.I., Carter, B.R., Saunders, M.P., Maranas, C.D., Woodward, M.J., and Anjum, M.F. (2009). Genome scale reconstruction of a *Salmonella* metabolic model: comparison of similarity and differences with a commensal *Escherichia coli* strain. J. Biol. Chem. *284*, 29480–29488.

Antoniewicz, M.R., Kraynie, D.F., Laffend, L.A., Gonzalez-Lergier, J., Kelleher, J.K., and Stephanopoulos, G. (2007). Metabolic flux analysis in a nonstationary system: fed-batch fermentation of a high yielding strain of *E. coli* producing 1,3-propanediol. Metab. Eng *9*, 277–292.

Bacon, J., James, B.W., Wernisch, L., Williams, A., Morley, K.A., Hatch, G.J., Mangan, J.A., Hinds, J., Stoker, N.G., Butcher, P.D., *et al.* (2004). The influence of reduced oxygen availability on pathogenicity and gene expression in *Mycobacterium tuberculosis*. Tuberculosis *84*, 205–217.

Bacon, J., Dover, L.G., Hatch, K.A., Zhang, Y., Gomes, J.M., Kendall, S., Wernisch, L., Stoker, N.G., Butcher, P.D., Besra, G.S., *et al.* (2007). Lipid composition and transcriptional response of *Mycobacterium tuberculosis* grown under iron-limitation in continuous culture: identification of a novel wax ester. Microbiology *153*, 1435–1444.

Baughn, A.D., Garforth, S.J., Vilcheze, C., and Jacobs, W.R., Jr. (2009). An anaerobic-type alpha-ketoglutarate ferredoxin oxidoreductase completes the oxidative tricarboxylic acid cycle of *Mycobacterium tuberculosis*. PLoS Pathog. *5*, e1000662.

Beste, D.J., Peters, J., Hooper, T., Avignone-Rossa, C., Bushell, M.E., and McFadden, J. (2005). Compiling a molecular inventory for *Mycobacterium bovis* BCG at two growth rates: evidence for growth rate-mediated regulation of ribosome biosynthesis and lipid metabolism. J. Bacteriol. *187*, 1677–1684.

Beste, D.J., Hooper, T., Stewart, G.S., Bonde, B., Avignone-Rossa, C., Bushell, M., Wheeler, P.R., Klamt, S., Kierzek, A.M., and McFadden, J.J. (2007a). GSMN-TB: a web-based genome scale network model of *Mycobacterium tuberculosis* metabolism. Genome Biol *8*, R89.

Beste, D.J., Laing, E., Bonde, B., Avignone-Rossa, C., Bushell, M.E., and McFadden, J.J. (2007b). Transcriptomic analysis identifies growth rate modulation as a component of the adaptation of mycobacteria to survival inside the macrophage. J. Bacteriol. *189*, 3969–3976.

Beste, D.J., Espasa, M., Bonde, B., Kierzek, A.M., Stewart, G.R., and McFadden, J. (2009). The genetic requirements for fast and slow growth in mycobacteria. PLoS. One *4*, e5349.

Beste, D.J., Bonde, B., Hawkins, N., Ward, J.L., Beale, M.H., Noack, S., Noh, K., Kruger, N.J., Ratcliffe, R.G., and McFadden, J. (2011). ^{13}C Metabolic flux analysis identifies an unusual route for pyruvate dissimilation in Mycobacteria which requires isocitrate lyase and carbon dioxide fixation. PLoS Pathog. *7*, e1002091.

Bettenbrock, K., Fischer, S., Kremling, A., Jahreis, K., Sauter, T., and Gilles, E.D. (2006). A quantitative approach to catabolite repression in *Escherichia coli*. J. Biol. Chem. *281*, 2578–2584.

Boer, V.M., de Winde, J.H., Pronk, J.T., and Piper, M.D. (2003). The genome-wide transcriptional responses of *Saccharomyces cerevisiae* grown on glucose in aerobic chemostat cultures limited for carbon, nitrogen, phosphorus, or sulfur. J. Biol. Chem. *278*, 3265–3274.

Boer, V.M., Tai, S.L., Vuralhan, Z., Arifin, Y., Walsh, M.C., Piper, M.D., de Winde, J.H., Pronk, J.T., and Daran, J.M. (2007). Transcriptional responses of *Saccharomyces cerevisiae* to preferred and nonpreferred nitrogen sources in glucose-limited chemostat cultures. FEMS Yeast Res. *7*, 604–620.

Bonde, B.K., Beste, D., Laing, E., Kierzek, A., and McFadden, J. (2011). Differential Producibility Analysis (DPA) of transcriptomic data with metabolic networks: Deconstructing the metabolic response of *M. tuberculosis*. PLoS Comput. Biol. *7*, e1002060.

Bordbar, A., Lewis, N.E., Schellenberger, J., Palsson, B.O., and Jamshidi, N. (2010). Insight into human alveolar macrophage and *M. tuberculosis* interactions via metabolic reconstructions. Mol. Syst. Biol. *6*, 422.

Boshoff, H.I., Myers, T.G., Copp, B.R., McNeil, M.R., Wilson, M.A., and Barry, C.E., III (2004). The transcriptional responses of *Mycobacterium tuberculosis* to inhibitors of metabolism: novel insights into drug mechanisms of action. J. Biol. Chem. *279*, 40174–40184.

Boyle, N.R., and Morgan, J.A. (2009). Flux balance analysis of primary metabolism in *Chlamydomonas reinhardtii*. BMC. Syst. Biol. *3*, 4.

Bumann, D. (2009). System-level analysis of *Salmonella* metabolism during infection. Curr. Opin. Microbiol. *12*, 559–567.

Cappelli, G., Volpe, E., Grassi, M., Liseo, B., Colizzi, V., and Mariani, F. (2006). Profiling of *Mycobacterium tuberculosis* gene expression during human macrophage infection: upregulation of the alternative sigma factor G, a group of transcriptional regulators, and proteins with unknown function. Res. Microbiol. *157*, 445–455.

de Carvalho, L.P., Zhao, H., Dickinson, C.E., Arango, N.M., Lima, C.D., Fischer, S.M., Ouerfelli, O., Nathan, C., and Rhee, K.Y. (2010). Activity-based metabolomic profiling of enzymatic function: identification of Rv1248c as a mycobacterial 2-hydroxy-3-oxoadipate synthase. Chem. Biol. *17*, 323–332.

Colijn, C., Brandes, A., Zucker, J., Lun, D.S., Weiner, B., Farhat, M.R., Cheng,T.Y., Moody, D.B., Murray, M., and Galagan, J.E. (2009). Interpreting expression data with metabolic flux models: predicting *Mycobacterium tuberculosis* mycolic acid production. PLoS Comput. Biol. *5*, e1000489.

Covert, M.W., Knight, E.M., Reed, J.L., Herrgard, M.J., and Palsson, B.O. (2004). Integrating high-throughput and computational data elucidates bacterial networks. Nature *429*, 92–96.

Daran-Lapujade, P., Daran, J.M., Kotter, P., Petit, T., Piper, M.D., and Pronk, J.T. (2003). Comparative genotyping of the *Saccharomyces cerevisiae* laboratory strains S288C and CEN.PK113-7D using oligonucleotide microarrays. FEMS Yeast Res. *4*, 259–269.

De Voss, J.J., Rutter, K., Schroeder, B.G., Su, H., Zhu, Y., and Barry,C.E., III. (2000). The salicylate-derived mycobactin siderophores of *Mycobacterium tuberculosis* are essential for growth in macrophages. Proc. Natl. Acad. Sci. U.S.A. *97*, 1252–1257.

Dias, M.V., Ely, F., Palma, M.S., de Azevedo, W.F.J., Basso, L.A., and Santos, D.S. (2007). Chorismate synthase: an attractive target for drug development against orphan diseases. Curr. Drug Targets *8*, 437–444.

Duarte, N.C., Becker, S.A., Jamshidi, N., Thiele, I., Mo, M.L., Vo, T.D., Srivas, R., and Palsson, B.O. (2007). Global reconstruction of the human metabolic network based on genomic and bibliomic data. Proc. Natl. Acad. Sci. U.S.A. *104*, 1777–1782.

Durot, M., Bourguignon, P.Y., and Schachter, V. (2009). Genome-scale models of bacterial metabolism: reconstruction and applications. FEMS Microbiol. Rev. *33*, 164–190.

Ely, F., Nunes, J.E., Schroeder, E.K., Frazzon, J., Palma, M.S., Santos, D.S., and Basso, L.A. (2008). The *Mycobacterium tuberculosis* Rv2540c DNA sequence encodes a bifunctional chorismate synthase. BMC Biochem. *9*, 13.

Fang, X., Wallqvist, A., and Reifman, J. (2009). A systems biology framework for modeling metabolic enzyme inhibition of *Mycobacterium tuberculosis*. BMC Syst. Biol. 3, 92.

Feist, A.M., and Palsson, B.O. (2008). The growing scope of applications of genome-scale metabolic reconstructions using *Escherichia coli*. Nat. Biotechnol. *26*, 659–667.

Feist, A.M., Henry, C.S., Reed, J.L., Krummenacker, M., Joyce, A.R., Karp, P.D., Broadbelt, L.J., Hatzimanikatis, V., and Palsson, B.O. (2007). A genome-scale metabolic reconstruction for *Escherichia coli* K-12 MG1655 that accounts for 1260 ORFs and thermodynamic information. Mol. Syst. Biol. 3, 121.

Feist, A.M., Herrgard, M.J., Thiele, I., Reed, J.L., and Palsson, B.O. (2009). Reconstruction of biochemical networks in microorganisms. Nat. Rev. Microbiol. 7, 129–143.

Fritz, C., Maass, S., Kreft, A., and Bange, F.C. (2002). Dependence of *Mycobacterium bovis* BCG on anaerobic nitrate reductase for persistence is tissue specific. Infect. Immun. *70*, 286–291.

Garton, N.J., Waddell, S.J., Sherratt, A.L., Lee, S.M., Smith, R.J., Senner, C., Hinds, J., Rajakumar, K., Adegbola, R.A., Besra, G.S., *et al.* (2008). Cytological and transcript analyses reveal fat and lazy persister-like bacilli in tuberculous sputum. PLoS. Med. *5*, e75.

Glickman, M.S., Cox, J.S., and Jacobs, W.R., Jr. (2000). A novel mycolic acid cyclopropane synthetase is required for cording, persistence, and virulence of *Mycobacterium tuberculosis*. Mol. Cell *5*, 717–727.

Hayes, A., Zhang, N., Wu, J., Butler, P.R., Hauser, N.C., Hoheisel, J.D., Lim, F.L., Sharrocks, A.D., and Oliver, S.G. (2002). Hybridization array technology coupled with chemostat culture: Tools to interrogate gene expression in *Saccharomyces cerevisiae*. Methods *26*, 281–290.

Heinrich, R., and Rapoport, T.A. (1974). A linear steady-state treatment of enzymatic chains. Critique of the crossover theorem and a general procedure to identify interaction sites with an effector. Eur. J. Biochem. *42*, 97–105.

Hoskisson, P.A., and Hobbs, G. (2005). Continuous culture – making a comeback? Microbiology *151*, 3153-3159.

James, B.W., Williams, A., and Marsh, P.D. (2000). The physiology and pathogenicity of *Mycobacterium tuberculosis* grown under controlled conditions in a defined medium. J. Appl. Microbiol. *88*, 669–677.

Jamshidi, N., and Palsson, B.O. (2007). Investigating the metabolic capabilities of *Mycobacterium tuberculosis* H37Rv using the in silico strain iNJ661 and proposing alternative drug targets. BMC. Syst. Biol. *1*, 26.

Jenkins, C., Bacon, J., Allnutt, J., Hatch, K.A., Bose, A., O'Sullivan, D.M., Arnold, C., Gillespie, S.H., and McHugh, T.D. (2009). Enhanced heterogeneity of rpoB in *Mycobacterium tuberculosis* found at low pH. J. Antimicrob. Chemother. *63*, 1118–1120.

Kacser, H., and Burns, J.A. (1995). The control of flux. Biochem. Soc. Trans. *23*, 341–366.

Kayser, A., Weber, J., Hecht, V., and Rinas, U. (2005). Metabolic flux analysis of *Escherichia coli* in glucose-limited continuous culture. I. Growth-rate-dependent metabolic efficiency at steady state. Microbiology *151*, 693–706.

Keating, L.A., Wheeler, P.R., Mansoor, H., Inwald, J.K., Dale, J., Hewinson, R.G., and Gordon, S.V. (2005). The pyruvate requirement of some members of the *Mycobacterium tuberculosis* complex is due to an inactive pyruvate kinase: implications for *in vivo* growth. Mol. Microbiol. *56*, 163–174.

Kim, H.U., Kim, T.Y., and Lee, S.Y. (2008). Metabolic flux analysis and metabolic engineering of microorganisms. Mol. Biosyst. *4*, 113–120.

Kim, T.Y., Kim, H.U., and Lee, S.Y. (2010). Metabolite-centric approaches for the discovery of antibacterials using genome-scale metabolic networks. Metab. Eng. *12*, 105–111.

Kitano, H. (2007). Towards a theory of biological robustness. Mol. Syst. Biol. *3*, 137.

Lyon, R.H., Hall, W.H., and Costas-Martinez, C. (1969). Uptake and distribution of labeled carbon from 14C-asparagine by *Mycobacterium tuberculosis*. J. Bacteriol. *98*, 317–318.

McKinney, J.D., Honer zu Bentrup, K., Munoz-Elias, E.J., Miczak, A.C.B., Chan, W.-T., Senson, D., Sacchettini, J.C., Jacobs-WR, J., and Russell, D.G. (2000). Persistance of *Mycobacterium tuberculosis* in macrophages and mice requires the glyoxylate shunt enzyme isocitrate lyase. Nature *406*, 735–738.

Miner, M.D., Chang, J.C., Pandey, A.K., Sassetti, C.M., and Sherman, D.R. (2009). Role of cholesterol in *Mycobacterium tuberculosis* infection. Indian J. Exp. Biol. *47*, 407–411.

Mo, M.L., Jamshidi, N., and Palsson, B.O. (2007). A genome-scale, constraint-based approach to systems biology of human metabolism. Mol. Biosyst. *3*, 598–603.

Movahedzadeh, F., Smith, D.A., Norman, R.A., Dinadayala, P., Murray-Rust, J., Russell, D.G., Kendall, S.L., Rison, S.C., McAlister, M.S., Bancroft, G.J., *et al.* (2004). The *Mycobacterium tuberculosis ino1* gene is essential for growth and virulence. Mol. Microbiol. *51*, 1003–1014.

Munoz-Elias, E.J., and McKinney, J.D. (2005). *Mycobacterium tuberculosis* isocitrate lyases 1 and 2 are jointly required for *in vivo* growth and virulence. Nat. Med. *11*, 638–644.

Munoz-Elias, E.J., and McKinney, J.D. (2006). Carbon metabolism of intracellular bacteria. Cell Microbiol. *8*, 10–22.

Munoz-Elias, E.J., Upton, A.M., Cherian, J., and McKinney, J.D. (2006). Role of the methylcitrate cycle in *Mycobacterium tuberculosis* metabolism, intracellular growth, and virulence. Mol. Microbiol. *60*, 1109–1122.

Oberhardt, M.A., Puchalka, J., Fryer, K.E., Martins dos Santos, V.A., and Papin, J.A. (2008). Genome-scale metabolic network analysis of the opportunistic pathogen *Pseudomonas aeruginosa* PAO1. J. Bacteriol. *190*, 2790–2803.

Oberhardt, M.A., Chavali, A.K., and Papin, J.A. (2009a). Flux balance analysis: interrogating genome-scale metabolic networks. Methods Mol. Biol. *500*, 61–80.

Oberhardt, M.A., Palsson, B.O., and Papin, J.A. (2009b). Applications of genome-scale metabolic reconstructions. Mol. Syst. Biol. *5*, 320.

Orth, J.D., Thiele, I., and Palsson, B.O. (2010). What is flux balance analysis? Nat. Biotechnol. *28*, 245–248.

Patil, K.R., and Nielsen, J. (2005). Uncovering transcriptional regulation of metabolism by using metabolic network topology. Proc. Natl. Acad. Sci. U.S.A. *102*, 2685–2689.

Peng, L., Arauzo-Bravo, M.J., and Shimizu, K. (2004). Metabolic flux analysis for a ppc mutant *Escherichia coli* based on 13C-labelling experiments together with enzyme activity assays and intracellular metabolite measurements. FEMS Microbiol. Lett. *235*, 17–23.

Raju, B., Hoshino, Y., Belitskaya-Levy, I., Dawson, R., Ress, S., Gold, J.A., Condos, R., Pine, R., Brown, S., Nolan, A., Rom, W.N., and Weiden, M.D. (2008). Gene expression profiles of bronchoalveolar cells in pulmonary TB. Tuberculosis *88*, 39–51.

Raman, K., and Chandra, N. (2009). Flux balance analysis of biological systems: applications and challenges. Brief. Bioinform. *10*, 435–449.

Raman, K., Rajagopalan, P., and Chandra, N. (2005). Flux balance analysis of mycolic acid pathway: targets for anti-tubercular drugs. PLoS Comput. Biol. *1*, e46.

Raman, K., Yeturu, K., and Chandra, N. (2008). targetTB: a target identification pipeline for *Mycobacterium tuberculosis* through an interactome, reactome and genome-scale structural analysis. BMC. Syst. Biol. *2*, 109.

Raman, K., Vashisht, R., and Chandra, N. (2009). Strategies for efficient disruption of metabolism in *Mycobacterium tuberculosis* from network analysis. Mol. Biosyst. *5*, 1740–1751.

Salmon, K., Hung, S.P., Mekjian, K., Baldi, P., Hatfield, G.W., and Gunsalus, R.P. (2003). Global gene expression profiling in *Escherichia coli* K12. The effects of oxygen availability and FNR. J. Biol. Chem. *278*, 29837–29855.

Sassetti, C.M., and Rubin, E.J. (2003). Genetic requirements for mycobacterial survival during infection. Proc. Natl. Acad. Sci. U.S.A. *100*, 12989–12994.

Sassetti, C.M., Boyd, D.H., and Rubin, E.J. (2003). Genes required for mycobacterial growth defined by high density mutagenesis. Mol. Microbiol. *48*, 77–84.

Schilling, C.H., and Palsson, B.O. (1998). The underlying pathway structure of biochemical reaction networks. Proc. Natl. Acad. Sci. U.S.A. *95*, 4193–4198.

Schnappinger, D., Ehrt, S., Voskuil, M.I., Liu, Y., Mangan, J.A., Monahan, I.M., Dolganov, G., Efron, B., Butcher, P.D., Nathan, C., *et al.* (2003). Transcriptional adaptation of *Mycobacterium tuberculosis* within macrophages: insights into the phagosomal environment. J. Exp. Med. *198*, 693–704.

Schuetz, R., Kuepfer, L., and Sauer, U. (2007). Systematic evaluation of objective functions for predicting intracellular fluxes in *Escherichia coli*. Mol. Syst. Biol. *3*, 119.

Shi, L., Sohaskey, C.D., Pfeiffer, C., Datta, P., Parks, M., McFadden, J., North, R.J., and Gennaro, M.L. (2010). Carbon flux rerouting during *Mycobacterium tuberculosis* growth arrest. Mol. Microbiol. *78*, 1199–1215.

Shlomi, T., Cabili, M.N., Herrgard, M.J., Palsson, B.O., and Ruppin, E. (2008). Network-based prediction of human tissue-specific metabolism. Nat. Biotechnol. *26*, 1003–1010.

Singh, A., Crossman, D.K., Mai, D., Guidry, L., Voskuil, M.I., Renfrow, M.B., and Steyn, A.J. (2009). *Mycobacterium tuberculosis* WhiB3 maintains redox homeostasis by regulating virulence lipid anabolism to modulate macrophage response. PLoS Pathog. *5*, e1000545.

Sohaskey, C.D., and Wayne, L.G. (2003). Role of *narK2X* and *narGHJI* in hypoxic upregulation of nitrate reduction by *Mycobacterium tuberculosis*. J. Bacteriol. *185*, 7247–7256.

Tailleux, L., Waddell, S.J., Pelizzola, M., Mortellaro, A., Withers, M., Tanne, A., Castagnoli, P.R., Gicquel, B., Stoker, N.G., Butcher, P.D., *et al.* (2008). Probing host pathogen cross-talk by transcriptional profiling of both *Mycobacterium tuberculosis* and infected human dendritic cells and macrophages. PLoS One *3*, e1403.

Talaat, A.M., Lyons, R., Howard, S.T., and Johnston, S.A. (2004). The temporal expression profile of *Mycobacterium tuberculosis* infection in mice. Proc. Natl. Acad. Sci. U.S.A. *101*, 4602–4607.

Thiele, I., Jamshidi, N., Fleming, R.M., and Palsson, B.O. (2009). Genome-scale reconstruction of *Escherichia coli*'s transcriptional and translational machinery: a knowledge base, its mathematical formulation, and its functional characterization. PLoS Comput. Biol. *5*, e1000312.

Tian, J., Bryk, R., Itoh, M., Suematsu, M., and Nathan, C. (2005). Variant tricarboxylic acid cycle in *Mycobacterium tuberculosis*: identification of alpha-ketoglutarate decarboxylase. Proc. Natl. Acad. Sci. U.S.A. *102*, 10670–10675.

Varma, A., and Palsson, B.O. (1994). Stoichiometric flux balance models quantitatively predict growth and metabolic by-product secretion in wild-type *Escherichia coli* W3110. Appl. Environ. Microbiol. *60*, 3724–3731.

Wayne, L.G. (1994). Cultivation of *Mycobacterium tuberculosis* for research purposes. In Tuberculosis: Pathogenesis, Protection and Control, Bloom, B.R., ed. (ASM Press, Washington DC).

Weber, I., Fritz, C., Ruttkowski, S., Kreft, A., and Bange, F.C. (2000). Anaerobic nitrate reductase (*narGHJI*) activity of *Mycobacterium bovis* BCG *in vitro* and its contribution to virulence in immunodeficient mice. Mol. Microbiol. 35, 1017–1025.

Wiechert,W. (2001). 13C metabolic flux analysis. Metab. Eng. 3, 195–206.

Sulfolobus Systems Biology: Cool Hot Design for Metabolic Pathways

8

Theresa Kouril, Alexey Kolodkin, Melanie Zaparty, Ralf Steuer, Peter Ruoff, Hans V. Westerhoff, Jacky Snoep, Bettina Siebers and the SulfoSYS consortium

Abstract

Life at high temperature challenges the stability of macromolecules and cellular components, but also the stability of metabolites, which has received little attention. For the cell, the thermal instability of metabolites means it has to deal with the loss of free energy and carbon, or in more extremes, it might result in the accumulation of dead-end compounds. In order to elucidate the requirements and principles of metabolism at high temperature, we used a comparative blueprint modelling approach of the lower part of the glycolysis cycle. The conversion of glyceraldehyde 3-phosphate to pyruvate from the thermoacidophilic Crenarchaeon *Sulfolobus solfataricus* P2 (optimal growth temperature 80°C) was modelled based on the available blueprint model of the eukaryotic model organism *Saccharomyces cerevisiae* (optimal growth temperature of 30°C). In *S. solfataricus* only one reaction is different, namely glyceraldehyde-3-phosphate is directly converted into 3-phosphoglycerate by the non-phosphorylating glyceraldehyde-3-phosphate dehydrogenase, omitting the extremely heat-instable 1,3-bisphosphoglycerate. By taking the temperature-dependent non-enzymatic (spontaneous) degradation of 1,3-bisphosphoglycerate in account, modelling reveals that a hot lifestyle requires a cool design.

Archaea

Archaea are fundamental components of biogeochemical cycles on earth and some are dominating in extreme environments (De Long 1998a, 2001; Chaban *et al.*, 2006). The assignment of Archaea as the third domain of life, in addition to Bacteria and Eukaryotes, based on the universal small subunit rRNA, was established in 1977 (Woese and Fox, 1977). Archaea are sometimes seen as a mosaic of bacterial and eukaryote life-forms, but they also exhibit numerous unique features. Bacterial-like features such as being unicellular, the lack of a nuclear membrane and organelles, the presence of circular DNA molecule, plasmids and the organization of genes in operon structures, are in contrast with mechanisms involved in information processing (e.g. transcription, translation), which represent less complex versions of the respective eukaryote equivalents.

However, Archaea differ in many aspects of their cell composition and molecular biology from Bacteria as well as from Eukaryotes. For example, instead of peptidoglycan (murein) in bacterial cell walls, in some methanogenic species pseudomurein is present (Konig *et al.*, 1982, 1983), while most other Archaea, consist of a paracrystalline surface layer (S-layer; König *et al.*, 2007) or harbour no cell wall at all (e.g. Thermoplasmatales; Itoh *et al.*, 2007).

In addition, Archaea are unique with respect to their metabolism, which is characterized by unusual, modified pathways. Beside the occurrence of special pathways like methanogenesis (for review see Thauer *et al.*, 2008), also some common pathways, like the oxidative pentose phosphate pathway found in Bacteria and Eukaryotes are absent from Archaea. Pentoses are either generated via the non-oxidative pentose phosphate pathway and/or the reversed ribulose monophosphate pathway (Orita *et al.*, 2005). Furthermore, the archaeal pathways are characterized by many, new and unusual enzymes, which substitute for classical bacterial and eukaryote counterparts (Siebers and Schönheit, 2005; Van der Oost and Siebers, 2007).

Life at high temperature

Although Archaea are ubiquitous and thrive in diverse habitats (De Long 1998b; De Long and Pace 2001) most Archaea that have been cultivated so far, are extremophiles, organisms that do not only tolerate, but require harsh environments for growth, e.g. low or high temperature, low or high pH, high salt concentrations or often combinations thereof.

Challenging habitats for prokaryotic growth are high-temperature environments, which comprise terrestrial and marine hot springs (e.g. geyser, black smokers, respectively). (Hyper)thermophiles grow in the range of 60°C up to 113°C (Stetter, 1999). One striking example is *Pyrolobus fumarii* with an optimal growth temperature of 106°C (growth range 90–113°C, pH 4.0–6.5; Blöchl *et al.*, 1997). Life at high temperature requires efficient adaptation strategies and in the past years a major focus has been the structural and mechanistic adaptation of cell components and macromolecules such as DNA, proteins and membranes (for review see Stetter 1999, Daniel and Cowan, 2000; Kawashima *et al.*, 2000). (Hyper) thermophilic organisms have evolved different strategies to adapt to environmental changes and beside intrinsic factors, which are determined within the macromolecules itself, extrinsic factors like the accumulation of compatible solutes, play a major role. Compatible solutes, such as amino acids (proline, glutamate, glycine), sugars (sucrose, trehalose), polyols and their derivates (Empadinhas *et al.*, 2006), can be accumulated up to high intracellular concentrations without interfering with metabolism. They protect against various stresses, such as desiccation, dehydration and osmotic, oxidative or temperature stress. One well-known example, detected in all three domains of life, is the disaccharide trehalose, which serves for example as osmolyte in order to balance changes in osmotic pressure and stabilizes proteins and membranes at high and low temperatures (for review see Elbein *et al.*, 2003). However, whereas the stability of macromolecules has been well addressed in the past, only minor attention has been paid to the effect of temperature on metabolites.

The model organism *Sulfolobus solfataricus* (strain P2, DSM1617; Zillig *et al.*, 1980) is an acidophilic thermophile, which belongs to the order of Sulfolobales within the Crenarchaeota. *S. solfataricus* was isolated from Solfatara at the Pisciarelli fumarole field in Naples, Italy and grows optimally at 80°C (60–92°C) and pH 3 (pH 2–4). It qualifies as an archaeal model organism for systems biology analyses, because it is genetically tractable and a well-characterized archaeal species (Worthington *et al.* 2003; Albers *et al.* 2006; Deng *et al.* 2009; Wagner *et al.* 2009). Furthermore, it is very easy to cultivate under defined conditions and grows heterotrophically with many different carbohydrates, such as monosaccharides (e.g. arabinose, glucose, galactose), disaccharides (e.g. maltose, sucrose) and polysaccharides (e.g. starch), as well as with amino acids or peptides (Grogan, 1989). Therefore, it qualifies for studying archaeal metabolic pathways. In addition, comprehensive microbiological, biochemical and functional genomics data have been assembled over the past 25 years and effective standard operating procedures have been established recently (Zaparty *et al.*, 2010).

Archaeal carbohydrate metabolism

Polysaccharides are a major source of carbon in the three domains of life. Their utilization generally involves extracellular hydrolysis, uptake of oligosaccharides by specific transporters and their intracellular hydrolysis to generate hexoses, like glucose, galactose, mannose and fructose. Subsequently, these monosaccharides are oxidized via well-conserved central metabolic pathways.

The archaeal carbohydrate metabolic network (CMN) is as complex as that of Bacteria and primitive Eukaryotes, but is characterized by many novel enzymes and biochemical pathways. Even the basic pathways of sugar degradation, like the Embden–Meyerhof–Parnas (EMP) or the Entner–Doudoroff (ED) pathway have undergone slight modifications in these organisms compared with the 'classic' pathways in Bacteria and Eukaryotes (Selig *et al.*, 1997; Ronimus *et al.*, 2003; Verhees *et al.*, 2003; Siebers and Schönheit, 2005; Van der Oost and Siebers, 2007). Many of the archaeal enzymes lack similarity with their bacterial and eukaryote counterparts, or are members of different, 'new' enzyme families (e.g. ADP/ATP-dependent hexo(gluco)kinases, phosphoglucose isomerase; Hansen *et al.*, 2003; Siebers and Schönheit, 2005). Interestingly, despite the utilization of new enzymes most of the intermediates known so far (e.g. in the EMP pathway) are conserved, highlighting the major impact of thermodynamic constraints on metabolic reactions.

However, so far only very limited information is available for utilization and growth on alternative carbon sources in Archaea (e.g. pentoses, sugar acids, aldehydes) and therefore, reconstruction of the archaeal carbon metabolic network (CMN) is incomplete and regulatory mechanisms such as nutritional adaptation or carbon homeostasis, as well as energetics are still far from being understood.

Metabolic network reconstruction

A prerequisite for systems biology is a complete network reconstruction in order to identify all players of the investigated network. For the *Sulfolobus* Systems Biology (SulfoSYS; www.sulfosys.com) project, the central carbohydrate metabolism (including the EMP and branched ED pathway, trehalose and glycogen metabolism) of *S. solfataricus* P2 has been reconstructed and the effect of temperature change on the metabolic network has been investigated (Albers *et al.*, 2009; Zaparty *et al.*, 2010).

The established *S. solfataricus* CMN blueprint model was used recently for pathway reconstruction within about 30 thermoacidophilic Archaea, distributed among the archaeal kingdoms of Crenarchaeota and Euryarchaeota, with available genome sequence information (Auernik *et al.*, 2008; Zaparty and Siebers, 2011).

Previous studies revealed that *S. solfataricus* uses an unusual branched ED pathway for the catabolism of glucose and galactose to pyruvate (Lamble *et al.*, 2005; Ahmed *et al.*, 2005). Because of the absence of a phosphofructokinase (PFK) homologue in *S. solfataricus*, it is suggested that the EMP pathway is utilized only in the gluconeogenic direction. The promiscuous ED pathway consists of a semi-phosphorylative (sp) and non-phosphorylative (np) branch (Fig. 8.1). Here, phosphorylation occurs via 2-keto-3-deoxygluconate kinase (KDGK) or glycerate kinase (GK), respectively (De Rosa *et al.*, 1984; Ahmed *et al.*, 2004, 2005; Lamble *et al.*, 2003, 2005).

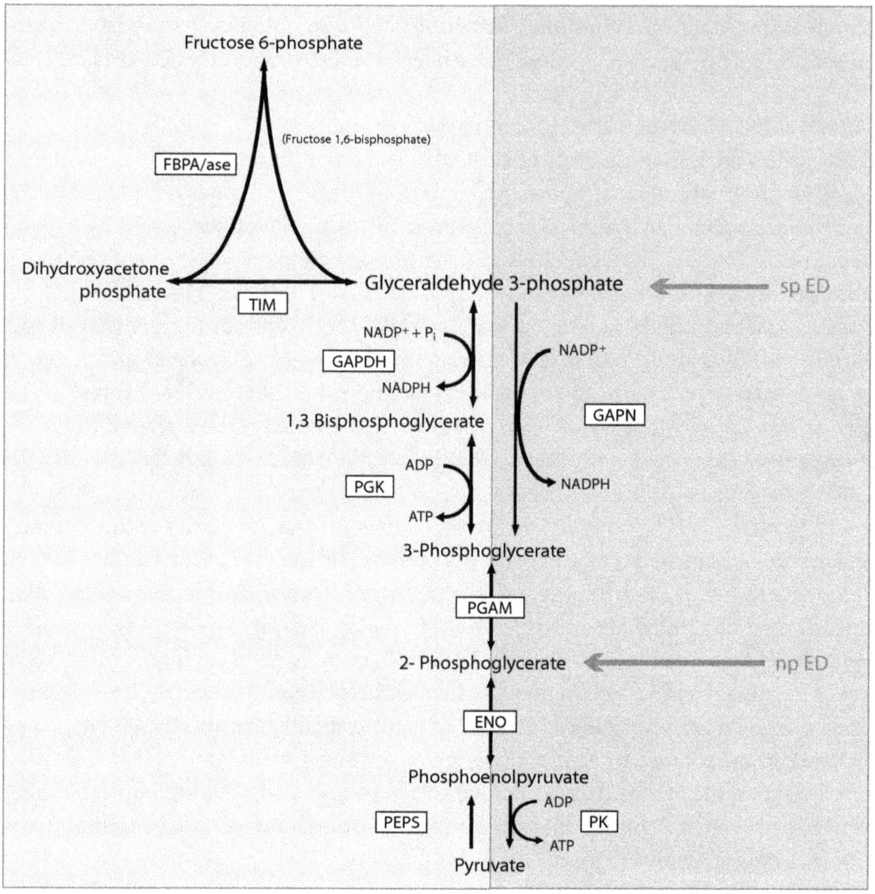

Figure 8.1 GAP conversion in *S. solfataricus.* Boxed in dark grey: reactions of the branched ED pathway; boxed in light grey: reactions of the anabolic EMP pathway. FBPA/ase, fructose-1,6-bisphosphate aldolase/phosphatase; TIM, triosephosphate isomerase; GAPDH, glyceraldehyde-3-phosphate dehydrogenase; GAPN, non-phosphorylating glyceraldehyde-3-phosphate dehydrogenase; PGK, phosphoglycerate kinase; PGAM, phosphoglycerate mutase; ENO, enolase; PK, pyruvate kinase; PEPS, phosphoenolpyruvate synthetase.

The common part of the branched ED pathway in *S. solfataricus* involves the initial oxidation of glucose/galactose to gluconate/galactonate by glucose dehydrogenase (GDH; Lamble *et al.*, 2003; Haferkamp *et al.*, 2011) and the dehydration by gluconate/galactonate dehydratase (GAD; Lamble *et al.*, 2004; Kim and Lee, 2005) forming 2-keto-3-deoxygluconate/2-keto-3-deoxygalactonate (KDG/KDGal), the characteristic intermediates of this pathway. At the beginning of the sp ED branch KDG/KDGal is phosporylated to 2-keto-3-desoxy-6-phosphogluconate/2-keto-3- deoxy-6-phosphogalactonate (KDPG/KDPGal) via KDGK. KDPG/KDPGal as well as KDG/KDGal are cleaved by the bifunctional KD(P)G aldolase, yielding glyceraldehyde 3-phosphate (GAP) and pyruvate in the sp branch and glyceraldehydes (GA) and pyruvate in the np route, respectively (Buchanan *et al.*, 1999; Lamble *et al.*, 2003, 2005; Ahmed *et al.*, 2005). GAP is further metabolized either via the glyceraldehyde-3-phosphate dehydrogenase (GAPDH)/phosphoglycerate kinase (PGK) couple (Hess *et al.*, 1995; Jones *et al.*, 1995; Russo *et al.*, 1995) or via the non-phosphorylating GAPDH (GAPN; Ettema *et al.*, 2008) ending in the lower common shunt of the EMP pathway, yielding a second molecule of pyruvate (Fig. 8.1).

In the np branch, GA is further oxidized to glycerate via glyceraldehyde dehydrogenase (GADH) or glyceraldehyde-oxidoreductase (GAOR), the latter has been characterized from *S. acidocaldarius* (aldehyde ferrodoxin oxidoreductase (AOR), Kardinahl *et al.*, 1999). Glycerate is then transformed to 2-phosphoglycerate (2PG) via glycerate kinase (GK) and channelled into the EMP pathway forming a second molecule of pyruvate (Zaparty and Siebers, 2011).

From GAP to pyruvate: the common shunt of the EMP pathway

A general feature of all (hyper)thermophilic Archaea analysed so far, is that they lack the characteristic bacterial and eukaryote regulation points at the beginning and end of the glycolytic EMP pathway. The archaeal ATP-dependent hexokinase, ADP-dependent glucokinase, ATP-, ADP-, and PP_i-dependent phosphofructokinases all lack allosteric properties, thus omitting the central control points of the classical EMP pathway as found in Bacteria and Eukaryotes. Also the archaeal pyruvate kinases (PK) exhibit reduced, if at all, regulatory potential (Siebers und Schönheit, 2005).

In *S. solfataricus* the upper part of the EMP pathway is only used for gluconeogenesis and the key metabolite at the branch point between gluconeogenic EMP and the lower glycolytic sp ED shunt of *S. solfataricus* is GAP (Fig. 8.1). In the anabolic direction GAP is converted to dihydroxyacetone phosphate (DHAP) by triosephosphate isomerase (TIM) and further metabolized to fructose 6-phosphate via an unusual, bifunctional fructose *1*,6-bisphosphate (FBP) aldolase/phosphatase (FBPA/ase, T. Kouril *et al.*, unpublished; Fig. 8.1). The FBPA/ase catalyses the direct conversion of DHAP and GAP to fructose 6-phosphate, as recently described for the hyperthermophilic Archaeon *Ignicoccus hospitalis* (Say *et al.*, 2010). This bifunctional enzyme seems to be restricted to Archaea and deep branching, mostly thermophilic and autothrophic Bacteria (Say *et al.*, 2010). However, bioinformatic analyses revealed, in addition, the presence of the archaeal-type class I FBP aldolase (Siebers *et al.*, 2001; Lorentzen *et al.*, 2003, 2005) homologue in *S. solfataricus*, which might catalyse reversible FBP formation. However, its role still has to be elucidated.

For the catabolic conversion of GAP, two different enzyme activities are known in *Sulfolobus*: (i) the classical GAPDH/PGK enzyme couple (Fabry *et al.*, 1987; Zwickel *et al.*, 1990; Jones *et al.*, 1995; Hess *et al.*, 1995; Russo *et al.*, 1995) and (ii) the GAPN (Brunner *et al.*, 1998, 2001; Ettema *et al.*, 2008), which catalyses the unidirectional oxidation of GAP to 3-phosphoglycerate (3PG; Fig. 8.1). In other Archaea, a third enzyme, the GAP oxidoreductase (GAP-OR), which catalyses the same reaction as GAPN, but uses ferredoxin instead of pyridine nucleotides as a co-substrate (Mukund *et al.*, 1995; Ettema *et al.*, 2008), has been identified. The formed 3PG is further converted to 2-phosphoglycerate (2PG), phosphoenolpyruvate (PEP) via phosphoglycerate mutase (PGAM) and enolase (ENO). Finally, a second molecule of pyruvate is formed via the pyruvate kinase (PK) reaction (Fig. 8.1).

Owing to its allosteric regulation, GAPN is the first and only regulation point of the sp ED pathway identified to date (Brunnner *et al.*, 1998, 2001; Lorentzen *et al.*, 2004; Ahmed *et al.*, 2005; Ettema *et al.*, 2008). The enzyme is efficiently activated by glucose 1-phosphate (G1P), a glycogen metabolism intermediate. Bioinformatics analyses revealed that GAPN seems to be present only in Archaea with an optimal growth temperature above 75°C, with the only exception of *Halobacterium* sp. NRC-1, a halophilic Archaeon with an optimal growth temperature of 37°C (Ettema *et al.*, 2008). In the catabolic direction, the use of GAPN instead of the GAPDH/PGK couple omits the extremely thermolabile intermediate 1,3-bisphosphoglycerate (1,3-BPG; half-life of 96 seconds at 60°C; Schramm *et al.*, 2001), therefore, a function of GAPN in metabolic thermoadaptation has been proposed previously (Ettema *et al.*, 2008). Detailed biochemical and transcriptional analyses in *Pyrococcus furiosus* and *Thermoproteus tenax* revealed a catabolic role for the GAPN and GAPOR, and an anabolic role for the GAPDH/PGK couple (Schäfer and Schönheit, 1993; Brunner *et al.*, 1998, 2001; Van der Oost *et al.*, 1998; Schut *et al.*, 2003; Lorentzen *et al.*, 2004).

However, the regulation at GAP level in *S. solfataricus* and other Archaea is still unclear and requires further biochemical investigations. The net ATP yield of the sp ED pathway is 1 mol ATP per mol of glucose, if the classical GAPDH/PGK couple is used, whereas no net energy is gained (0 mol ATP/mol glucose), if the GAPN or the np branch are used in glycolysis.

Blueprint modelling: glycolysis of *S. solfataricus* and *S. cerevisiae* are instantiations of the same master theme

Different biological organisms might be quite similar in a very important underlying mechanism of their function. First, building blocks (biomolecules) are basically the same. If L-amino acids and D-sugars had been chosen to be synthesized by the first ancestor organism, perhaps randomly, the evolution of successors would continue to work with those forms. If, at a certain stage, autotrophs have centred their metabolism on glucose, heterotrophs would be constrained to arrange their metabolism in a compatible way, to have compatible building blocks. We may also expect that major pathways, major biochemical reaction networks would remain similar through all organisms. First of all, the number of biologically affordable reactions is limited. We do not expect that any organism might ever synthesize glucose from water and CO_2 directly, in a single reaction; neither, would we expect that glucose would be oxidized to CO_2 and water directly. On the contrary, we would anticipate to find a thermodynamically reasonable, stepwise process compatible with life, which neither delivers nor consumes too much of Gibbs free energy. Since already emerged; most probably, the core scheme of this stepwise process would be inherited further and with the evolution of a biosphere as a whole; single organisms (species) would be constrained to stick to this consensus mechanism.

This seems to be also true for Archaea, although they use modifications of the classical EMP and ED pathway, most metabolites and catalysed reactions are similar, only the catalysts involved change. Similarity of biochemical pathways in various organisms justify a blueprint modelling approach: a blueprint (core) model may be built and then parameterized for a pathway of a particular organism, by varying only several parameters (assuming that the same kinetic type is used), such as the level of enzyme expression.

Here, we focus on a blueprint modelling of the lower part of glycolysis, the conversion of GAP to pyruvate. First, we project the reaction network diagram of this part of the glycolytic pathway in *S. solfataricus* over the reaction network of *S. cerevisiae*, an organism from the eukaryote kingdom, with an optimal growth at 30°C. The majority of the reactions overlap (Fig. 8.2) and only one reaction is different. As discussed above, apart from the classical GAPDH/PGK enzyme couple (reactions 1 and 2, Fig. 8.2), *S. solfataricus* employs

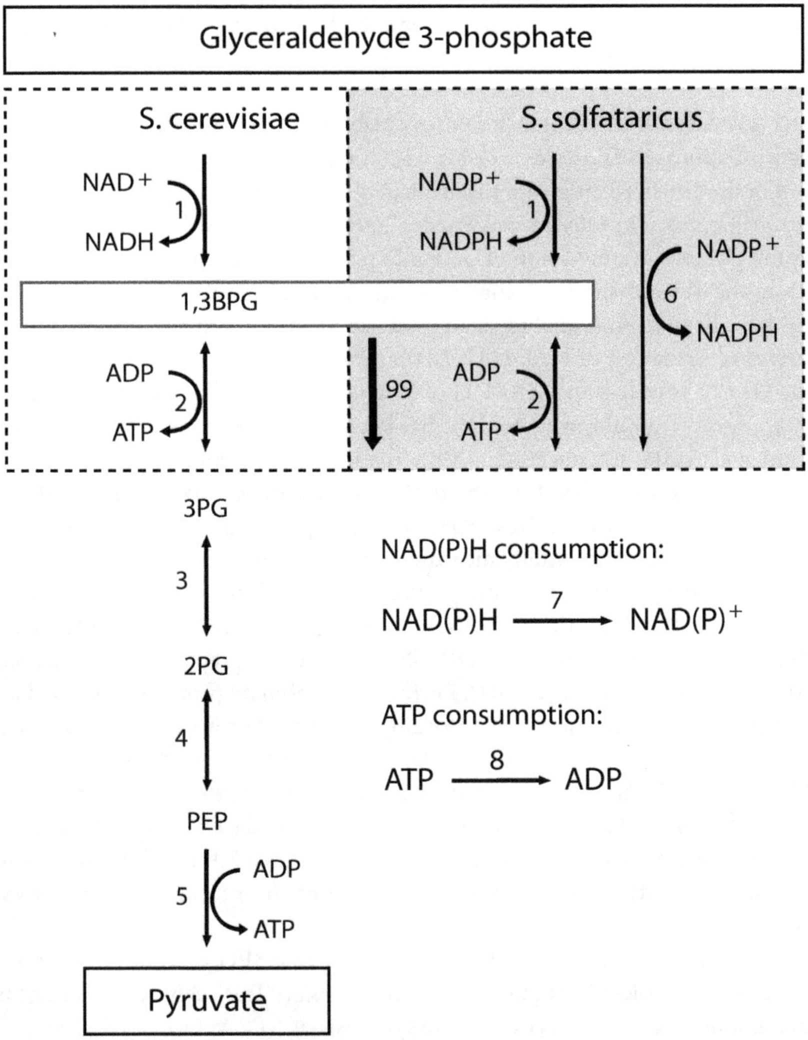

Figure 8.2 Blueprint scheme of the lower part of glycolysis (conversion of glyceraldehyde 3-phosphate to pyruvate) for *S. cerevisiae* and for *S. solfataricus*. GAP, glyceraldehyde-3-phosphate; 1,3BPG, 1,3-bisphosphoglycerate; 3PG, 3-phosphoglycerate; 2PG, 2-phosphoglycerate; PEP, phosphoenolpyruvate.

the GAPN, which catalyses the unidirectional oxidation of GAP to 3PG (reaction 6, Fig. 8.2; Ettema *et al.*, 2008), and the organism uses $NADP^+$ instead of NAD^+ as cofactor for these reactions. *S. cerevisiae* lacks GAPN and is unable to convert GAP into 3PG directly (Fig. 8.2). Consequently, *S. cerevisiae* glycolysis is, in fact, the glycolysis of *S. solfataricus*, but without reaction 6. This allows the building of a mathematical model of glycolysis in *S. solfataricus* by adding the extra reaction into the model of *S. cerevisiae* and re-parameterizing the kinetics of the other reactions.

Another factor that has to be considered is the temperature dependent non-enzymatic (spontaneous) degradation of 1,3-BPG (reaction 99, Fig. 8.2). For *S. cerevisiae*, this reaction may be neglected, but for *S. solfataricus* (living at high temperature) this has to be taken into account.

Building a mathematical model of glycolysis in *S. solfataricus*

The blueprint approach allows us to start with a well-known model of *S. cerevisiae* glycolysis, built in the Amsterdam group (Teusink *et al.*, 2000) and available at the JWS Online site (http://jjj.mib.ac.uk; Olivier and Snoep, 2004). Since we were interested only in the lower part of glycolysis (conversion of GAP into pyruvate), the corresponding part has been copied from the Teusink model. Initial concentrations of boundary metabolites (GAP and pyruvate) have been fixed (equal to steady state concentration in the Teusink model). The rate constant of oxidation of NADH (NADPH for *S. solfataricus*) has been assumed constant. The GAPN as well as the GAPDH of *S. solfataricus* have been shown to possess dual co-substrate specificity; however, $NADP^+$ has been shown to be the preferred co-substrate (Littlechild *et al.*, 2001; Ettema *et al.*, 2008). The rate of ATP consumption has been modelled with mass action kinetics. The ratio of the steady state ATP consumption flux and the GAP consumption flux indicate the efficiency of ATP production. Parameters for reactions 1, 5 and 6 have been experimentally measured within the SulfoSYS project (www.sulfosys.com). The specific activities (V_m values) were determined in cell-free extracts of *S. solfataricus* cells grown at 80°C [see SOP_SSO_080904 (fermenter set-up and fermentation); SOP_SSO_080905 (cell harvest); SOP_SSO_080913d (preparation of *S. solfataricus* crude extracts) in Zaparty *et al.* (2010)]. Purified, recombinant proteins were used to obtain Km values, performing enzyme assays according to standard protocols (Zaparty *et al.*, 2010; GAPN (reaction 6, Fig. 8.2) Ettema *et al.*, 2008; GAPDH (reaction 1, Fig. 8.2) Russo *et al.*, 1995; PK (reaction 5, Fig. 8.2) Schramm *et al.*, 2000). Parameters for reaction 2, 3 and 4 have been kept identical to the ones in the *S. cerevisiae* model (Teusink *et al.*, 2000). All rate equations, mass balances and parameters are shown in Table 8.1 and the model is available via JWS Online (e.g. http://jjj.mib.ac.uk, search e.g. for author Kouril), and can be exported in SBML format.

As shown in Fig. 8.2, when directly converting GAP into 3PG, *S. solfataricus* produces less ATP molecules per molecule of glucose than *S. cerevisiae*. This seems paradoxical, because the main function of glycolysis is exactly the synthesis of ATP. Why does *S. solfataricus* possess reaction 6? Would *S. solfataricus* not produce more ATP without this reaction?

The presence of GAPN might be understandable, if reactions 1 and 2 would be limiting for the total flux of glucose through the glycolytic pathway. However, if reactions 1 and 2 are sufficient for *S. cerevisiae*, living at low temperatures, the conversion of GAP into 3PG at high temperature should be even simpler due to the spontaneous degradation of 1,3-BPG. The non-enzymatic degradation of 1,3BPG would help to increase the total flux by stimulating reaction 1 (consuming its product) and by duplicating reaction 2. Another explanation for the existence of GAPN may be found, if there would be a high flux through reaction 99. Then the total mass flow would pass mostly through reactions 1 and 99 and there would be very little synthesis of ATP from GAP. Consequently, redirecting of the total flux from reactions 1 and 2 into reaction 6 would have almost no effect on the net outcome of ATP. Then, instead of producing GAPDH, *S. solfataricus* might produce GAPN. This maybe the reason why *Sulfolobus* possesses GAPN – simply because reaction 6 does not give any substantial disadvantages. However, it is not considered yet that 1,3-BPG might spontaneously degrade to compounds different from 3PG (e.g. 1PG) or other 'grey' metabolites. There are no studies about the decay products available so far. These 'grey' metabolites may simply escape from further steps of glycolysis and thus, may be lost from the point of ATP production (carbon loss). Therefore, it may be important to assure that the flux goes through GAPN. Then, *S. solfataricus* faces the dilemma of either to lose ATP, due to employing GAPN instead of PGK, or may be to lose metabolites from the glycolytic pathway. Is this the complete picture for the trade-off between those two processes?

Futile ATP cycling: inspirations from the model

The mathematical model for *S. solfataricus* allowed us to compare the *S. solfataricus* and *S. cerevisiae* GAP to pyruvate conversion as instantiations of the same blueprint master-scheme. To be able to compare the *Sulfolobus* and *Saccharomyces* systems we need to express the efficiency of the lower branch of glycolysis in terms of ATP produced per GAP consumed. In this way the lower efficiency of *Sulfolobus* due to its use of the ED pathway does not interfere with the estimations of the efficiency of the second half of glycolysis.

In *S. cerevisiae* at low temperature, neither reaction 99 nor reaction 6 are present, then two molecules of ATP will be produced per molecule of GAP converted to pyruvate. We expected a decrease in efficiency (ATP/GAP) after the addition of reaction 99 (spontaneous degradation of 1,3BPG to 3PG), since this could lead to 1,3BPG consumption uncoupled from ATP synthesis. We expected that at low temperature the effect of spontaneous 1,3BPG degradation would not be very severe, the flux through reaction 99 would be low and the ATP/GAP flux ratio would be somewhere between one and two. We also expected that at high temperature, the total pool of 1,3BPG would be degraded spontaneously, the flux through reaction 2 would approach zero and, consequently, the net ATP/GAP flux ratio would become equal to one. Model simulations confirmed these predictions for the case of low temperature. The ATP/GAP flux ratio was found to be between 1 and 2 (Table 8.2); but, at high temperature the ATP/GAP flux ratio dropped even further to values below 1. This phenomenon might be explained by futile cycling of ATP. The PGK reaction is simulated as a reversible reaction and the increased conversion of 1,3BPG due to degradation at high temperatures can lead to a change in 3PG/1,3BPG ratios and thereby increasing the reverse rate of PGK. The latter is coupled with the consumption of ATP and could be observed in the model as the negative flux value through reaction 2. The activity of the futile cycle reduced the ATP/GAP flux ratio below 1 and this indicates that removal of PGK could actually lead to a higher ATP/GAP flux ratio, although this would remove a potential ATP production site and limit the ratio to 1. Indeed, when reactions 1 and 2 in the *S. solfataricus* model were removed, the ATP/GAP flux ratio did not drop below 1, showing that this design was more advantageous.

Conclusions and future perspective

In the course of the SulfoSYS project (Albers *et al.*, 2009) a blueprint modelling approach with input of real kinetic data allowed us to address the design of a partial glycolysis model, in other words conversion of GAP to pyruvate for the thermoacidophilic archaeon *S. solfataricus.*

Metabolite instability is one important challenge for organisms growing at high temperature. Especially as some triosephosphates are heat-labile compounds, which need to be quickly converted to more heat-stabile carbohydrates in order to prevent carbon and energy

Table 8.1 Model description

Reactions		Parameters
v_1	GAPDH: GAP oxidation to 1,3-BPG (coupled with $NADP^+$ reduction) $\frac{\frac{V_{mfGAPDH}\cdot GAP(t)\cdot NADP(t)}{K_{mGAPDHGAP}\cdot K_{mGAPDHNADP}}-\frac{V_{mbGAPDH}\cdot BPG(t)\cdot NADPH(t)}{K_{mGAPDHBPG}\cdot K_{mGAPDHNADPH}}}{\left(1+\frac{GAP(t)}{K_{mGAPDHGAP}}+\frac{BPG(t)}{K_{mGAPDHNADP}}\right)\cdot\left(1+\frac{NADP(t)}{K_{mGAPDHBPG}}+\frac{NADPH(t)}{K_{mGAPDHNADPH}}\right)}$ (mmol/min)	$K_{mGAPDHGAP}$ = 4.6 mmol $K_{mGAPDHNADP}$ = 0. 31 mmol $K_{mGAPDHBPG}$ = 92.4 mmol $K_{mGAPDHNADPH}$ = 0.025 mmol *$V_{mfGAPDH}$ = $V_{mbGAPDH}$ = 0.066 U/mg = 66 mmol/min (SulfoSys data)
v_2	PGK (reversible): when in direction from BPG to 3PG generates ATP (Prb): V_{mPGK}*(K_{eqPGK}*BPG*(SUMAXP-(SUMAXP^2–2*SUMAXP*Prb+8*K_{eqAK}*SUMAXP*Prb+Prb^2–4*K_{eqAK}*Prb^2)^0.5)/(1–4*K_{eqAK})-(-SUMAXP+Prb-4*K_{eqAK}*Prb+(SUMAXP^2–2*SUMAXP*Prb+8*K_{eqAK}*SUMAXP*Prb+Prb^2–4*K_{eqAK}*Prb^2)^0.5)*3PG/(2–8*K_{eqAK}))/($K_{mPGKATP}$*$K_{mPGK3PG}$*(1+(SUMAXP-(SUMAXP^2–2*SUMAXP*Prb+8*K_{eqAK}*SUMAXP*Prb+Prb^2–4*K_{eqAK}*Prb^2)^0.5)/((1–4*K_{eqAK})*$K_{mPGKADP}$)+(–SUMAXP+Prb-4*K_{eqAK}*Prb+(SUMAXP^2–2*SUMAXP*Prb+8*K_{eqAK}*SUMAXP*Prb+Prb^2–4*K_{eqAK}*Prb^2)^0.5)/((2–8*K_{eqAK})*$K_{mPGKATP}$))*(1+BPG/$K_{mPGKBPG}$+3PG/$K_{mPGK3PG}$)) (mmol/min)	K_{eqAK} = 0.450 K_{eqPGK} = 3200 $K_{mPGKADP}$ = 0.2 mmol $K_{mPGKATP}$ = 0.3 mmol $K_{mPGKBPG}$ = 0.003 mmol $K_{mPGK3PG}$ = 0.53 mmol V_{mPGK} = 1306.45 mmol/min SUMAXP = 4.1 mmol As for yeast (Teusink *et al.*, 2000)
v_3	PGAM (reversible): 3PG to 2PG $\frac{\frac{V_{mPGAM}}{K_{mPGAM3PG}}\cdot 3PG(t)-\frac{2PG(t)}{K_{eqPGM}}}{1+\frac{3PG(t)}{K_{mPGAM3PG}}+\frac{2PG(t)}{K_{mPGAM2PG}}}$ (mmol/min)	V_{mPGAM} = 2525.81 mmol/min K_{eqPGAM} = 0.19 $K_{mPGAM3PG}$ = 1.2 mmol $K_{mPGAM2PG}$ = 0.08 mmol As for yeast (Teusink *et al.*, 2000)
v_4	ENO (reversible): 2PG to PEP $\frac{\frac{V_{mENO}}{K_{mENO2PG}}\cdot(2PG-\frac{PEP}{K_{eqENO}})}{1+\frac{2PG}{K_{mENO2PG}}+\frac{PEP}{K_{mENOPEP}}}$ (mmol/min)	V_{mENO}=365.8 mmol/min $K_{mENO2PG}$ = 0.04 mmol K_{eqENO} = 6.7 $K_{mENOPEP}$ = 0.5 mmol As for yeast (Teusink *et al.*, 2000)
v_5	PK (dephosphorylation of PEP to PYR, coupled with ATP (Prb) synthesis): V_{mPYK}/($K_{mPYKPEP}$*$K_{mPYKADP}$)*(PEP*(SUMAXP-(Prb^2–4*K_{eqAK}*Prb^2–2*Prb*SUMAXP+8*K_{eqAK}*Prb*SUMAXP+SUMAXP^2)^0.5)/(1–4*K_{eqAK})-PYR*((Prb-4*K_{eqAK}*Prb-SUMAXP+(Prb^2–4*K_{eqAK}*Prb^2–2*Prb*SUMAXP+8*K_{eqAK}*Prb*SUMAXP+SUMAXP^2)^0.5)/(2–8*K_{eqAK}))/K_{eqPYK})/((1+PEP/$K_{mPYKPEP}$+PYR/$K_{mPYKPYR}$))*(1+(Prb-4*K_{eqAK}*Prb-SUMAXP)+(Prb^2–4*K_{eqAK}*Prb^2–2*Prb*SUMAXP+8*K_{eqAK}*Prb*SUMAXP+SUMAXP^2)^0.5)/(2–8*K_{eqAK})/$K_{mPYKATP}$+(SUMAXP-(Prb^2–4*K_{eqAK}*Prb^2–2*Prb*SUMAXP+8*K_{eqAK}*Prb*SUMAXP+SUMAXP^2)^0.5)/(1–4*K_{eqAK})/($K_{mPYKADP}$) (mmol/min)	V_{mPYK}* = 0.04 U/mg = 40 mmol/min (measured for 70°C, for 80°C its value should be a bit higher) $K_{mPYKPEP}$ =0.26 mmol $K_{mPYKADP}$ =0.113 mmol (SulfoSys data) K_{eqAK} = 0.450 K_{eqPYK} = 6500 $K_{mPYKATP}$ =1.5 mmol $K_{mPYKPYR}$=21 mmol SUMAXP = 4.1 mmol As for yeast (Teusink *et al.*, 2000)

Reactions		Parameters
v_6	GAPN (irreversible): Oxidation of GAP to 3PG, coupled with $NADP^+$ reduction: $\frac{V_{mGAPN} \cdot GAP(t) \cdot NADP(t)}{K_{mGAPNGAP} \cdot NADP(t) + K_{mGAPNNADP} \cdot GAP(t) + GAP(t) \cdot NADP(t) + K_{mGAPNGAP} \cdot K_{mGAPNNADP}}$ (mmol/min)	$^*V_{mGAPN}$ =0.06 U/mg= 60 mmol/min $K_{mGAPNGAP}$ =1 mmol $K_{mGAPNNADP}$ =0.18 mmol (SulfoSys data)
v_7	Oxidation of NADPH+H^+ to $NADP^+$: $k_1 \cdot NADPH(t)$ (mmol/min)	k_1 = 170/min (fitted)
v_8	Consumption of ATP (Prb): $k_8 \cdot Prb(t)$ (mmol/min)	k_8 = 2.005 min^{-1} (fitted)
v_{99}	Spontaneous (non-enzymatic) degradation of BPG to 3PG: $k_{99} \cdot 3PG(t)$ (mmol/min)	k_{99} (varied) variable 10^5 and 10^7

Balance equations	
dNADP/d*t*	$(v_7 - v_6 - v_1)$ (mmol/min)
dNADPH/d*t*	$(v_6 + v_1 - v_7)$ (mmol/min])
d3PG/d*t*	$(v_{99} + v_{6+} v_2 - v_3)$ [mmol/min]
d2PG/d*t*	$(v_3 - v_4)$ (mmol/min)
dPEP/d*t*	$(v_4 - v_5)$ (mmol/min)
dBPG/d*t*	$(v_1 - v_2 - v_{99})$ (mmol/min)
dPYR/d*t*	(v_5) (mmol/min)
dTRIO/d*t*	$(-v_1 - v_6 - v_{99})$ (mmol/min)
dPrb/d*t* (=ATP)	$(v_2 + v_5 - v_8)$ (mmol/min)
Conserved moieties	
NADP(t) + NADPH(t)	1.58 (mmol)
Initial conditions	
Prb(0)	6.3 (mmol)
NADP(0)	1.54 (mmol) [1.48 mmol for design 4 (only GAPN)]
BGP(0)	0.0000522 (mmol)
NADPH(0)	0.04 (mmol) [0.1 mmol for design 4 (only GAPN)]
3PG(0)	0.08 (mmol) [1.28 mmol for design 4 (only GAPN)]
PEP(0)	0.09 (mmol) [1.37 mmol for design 4 (only GAPN)]
2PG(0)	0.01 (mmol) [0.23 mmol for design 4 (only GAPN)]
PYR(0)	(fixed) 8.5 (mmol)
GAP(0)	(fixed) 0.77 (mmol)
SUMAXP	(fixed) 4.1 (mmol)

GAP, glyceraldehyde 3-phosphate; BPG, 1,3-bisphosphoglycerate; 3PG, 3-phosphoglycerate; 2PG, 2-phosphoglycerate; PEP, phosphoenolpyruvate; PYR, pyruvate; Prb, corresponds to 2ATP + ADP, SUMAXP, corresponds to ADP, ATP and AMP, fixed.
$^*V_m$ was measured in U/mg of crude (total) protein (μmol/min/mg) and converted to (mmol/min) using the conversion factor 1 g of total protein corresponds to 1 ml of cell volume [parameter set in the realistic range, for yeast 1 g = 3.7 ml (Teusink *et al*, 2000), for mitochondria 1 g = 1 ml]

Table 8.2 Steady-state fluxes in (mmol/min) for *S. cerevisiae* and *S. solfataricus*. Mass balances, rate equations and parameters of corresponding models are shown in Table 8.1

	S. cerevisiae instantiation		*S. solfataricus* instantiation	
	Low temperature	High temperature (moderate rate of reaction 99)	Very high temperature (very high rate of reaction 99)	
	Reaction 1 Reaction 2	Reaction 1 Reaction 2 plus reaction 99 ($kf_{99} = 10^5$)	Reaction 1 Reaction 2 plus reaction 99 ($kf_{99} = 10^7$)	Reaction 6 only
ATP flux (ATP consumption)	12.6	10.14	1.82	25.6
GAP flux reaction v3 (v1 + v6)	6.3	6.3	6.3	25.6
ATP/GAP Flux ratio	2	1.61	0.3	1
Flux through reaction 99	0	2.48	10.8	–
Flux through reaction 2	6.3	3.82	-4.5	–
Flux through reaction 6	–	–	–	25.6

loss (Ahmed *et al.*, 2004; Ettema *et al.*, 2008; Say *et al.*, 2010). However, the decay products of the spontaneous degradation of *1*,3BPG are still unknown and part of current studies. If other 'grey' metabolites are formed, they may escape from glycolysis.

In both organisms, the mesophile *S. cerevisiae* and the thermophile *S. solfataricus*, GAP is produced via two different glycolytic pathways. *S. cerevisiae* metabolizes GAP via the classical EMP pathway, whereas in *S. solfataricus* GAP is formed via the sp ED branch; no GAP is produced in the np ED branch of the *S. solfataricus* pathway. In the model, the main focus lies on the ATP production in terms of ATP produced per molecule of GAP converted into pyruvate.

S. cerevisiae forms two ATP per molecule of GAP at low temperature, which might drop to below one ATP per molecule GAP at high temperature, by taking the temperature induced non-enzymatic degradation of 1,3BPG into account (futile cycling leads to a negative flux value of ATP through PGK reaction). The circumvention of 1,3BPG by the use of GAPN, as in *S. solfataricus*, results in the formation one ATP per GAP molecule.

Consequently it may be advantageous to remove GAPDH/PGK and involve GAPN or GAP-OR enzymes in (hyper)thermophiles, which catalyse the direct oxidation of GAP to 3PG. This fits with the above-mentioned bioinformatics analyses, showing that GAPN seems to be mostly present in Archaea with an optimal growth temperature above 75°C (Ettema *et al.*, 2008).

We have shown here that the GAPDH/PGK enzyme pair may cause a futile cycling: at high temperatures 1,3BPG would degrade spontaneously and it can be re-synthesized by reversed PGK activity resulting in the consumption of ATP. Thus, to circumvent the futile cycle it is essential to remove the PGK even more so than adding GAPN (or GAP-OR). However, GAPN cannot substitute entirely for the GAPDH/PGK couple. Both GAPDH and PGK are essential for gluconeogenesis.

With our blueprint modelling approach we have shown that the GAPDH/PGK route could lead to a net ATP consumption at high temperatures due to futile cycling over the PGK reaction. It could be shown that in this case it was better to utilize GAPN (in absence of PGK). If the net flux over the PGK is in the forward direction, this will lead to an ATP/GAP flux ratio higher than 1, and then the GAPDH/PGK route is more efficient than the GAPN route. The relative fluxes over the GAPDH/PGK, GAPN and spontaneous degradation of 1,3BPG at high temperature is dependent on the kinetics for these reactions, which are currently under experimental investigation.

Indeed, in previous studies we observed a tremendous activation of GAPN by G1P (intermediate in glycogen metabolism). The activity of GAPN increased about four-fold on addition of G1P (0.01 mmol; Ettema *et al.*, 2008), such an activation could lead to an increased flux over the GAPN route.

In summary, the 'hot' design of *S. solfataricus* does not seem to be a strange design wasting ATP compared with the 'cool' design of *S. cerevisiae* (Fig. 8.3), but rather representing a 'cool' adaptation strategy to high temperature. We present two possible mechanisms to explain why at high temperatures with an unstable 1,3BPG, the GAPN route may be advantageous over the GAPDH/PGK route for the conversion of GAP to 3PG. The first mechanism relates directly to the degradation of 1,3BPG which could lead to a carbon drain via 'grey' metabolites. The second mechanism relates to futile cycling leading to a net ATP drain via the PGK reaction.

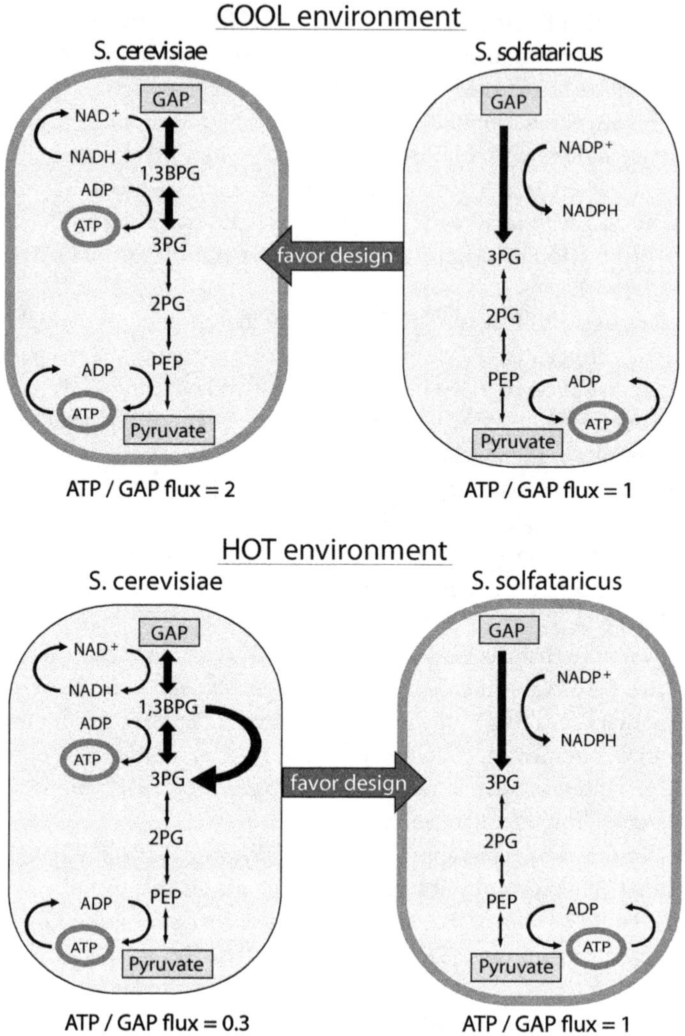

Figure 8.3 At low temperature, the conversion of GAP into 3PG via 1,3BPG is more preferable because two molecules of ATP may be generated per one molecule of GAP (ATP/GAP flux ratio = 2). At high temperature, ATP would be lost due to futile cycling: spontaneous degradation of 1,3BPG and re-synthesis of 1,3BPG from 3PG would consume ATP. Thus, at high temperature, it would be more preferable to omit 1,3BPG by the utilization of GAPN.

References

Ahmed, H., Tjaden, B., Hensel, R., and Siebers, B. (2004). Embden–Meyerhof–Parnas and Entner–Doudoroff pathways in *Thermoproteus tenax*: metabolic parallelism or specific adaptation? Biochem. Soc. Trans. 32, 303–304.

Ahmed, H., Ettema, T.J.G., Tjaden, B., Geerling, A.C.M., Van Der Oost, J., and Siebers, B. (2005). The semi-phosphorylative Entner–Doudoroff pathway in hyperthermophilic archaea: a re-evaluation. Biochem. J. *390*, 529–540.

Albers, S.V., Jonuscheit, M., Dinkelaker, S., Urich, T., Kletzin, A., Tampé, R., Driessen, A.J.M., and Schleper, C. (2006). Production of recombinant and tagged proteins in the hyperthermophilic archaeon *Sulfolobus solfataricus*. App. Environ. Microbiol. 72, 102–111.

Albers, S.V., Birkeland, N.K., Driessen, A.J.M., Gertig, S., Haferkamp, P., Klenk, H.P., Kouril, T., Manica, A., Pham, T.K., Ruoff, P., *et al.* (2009). SulfoSYS (Sulfolobus Systems Biology): towards a silicon cell model for the central carbohydrate metabolism of the archaeon *Sulfolobus solfataricus* under temperature variation. Biochem. Soc. Trans. *37*, 58–64.

Auernik, K.S., Cooper, C.R., and Kelly, R.M. (2008). Life in hot acid: pathway analyses in extremely thermoacidophilic archaea. Curr. Opin. Biotechnol. *19*, 445–453.

Blöchl, E., Rachel, R., Burggraf, S., Hafenbradl, D., Jannasch, H.W., and Stetter, K.O. (1997). *Pyrolobus fumarii*, gen. and sp. nov., represents a novel group of archaea, extending the upper temperature limit for life to 113°C. Extremophiles *1*, 14–21.

Brunner, N.A., Brinkmann, H., Siebers, B., and Hensel, R. (1998). NAD+-dependent glyceraldehyde-3-phosphate dehydrogenase from *Thermoproteus tenax*. The first identified archaeal member of the aldehyde dehydrogenase superfamily is a glycolytic enzyme with unusual regulatory properties. J. Biol. Chem. *273*, 6149–6156.

Brunner, N.A., Siebers, B., and Hensel, R. (2001). Role of two different glyceraldehyde-3-phosphate dehydrogenases in controlling the reversible Embden–Meyerhof–Parnas pathway in *Thermoproteus tenax*: Regulation on protein and transcript level. Extremophiles 5, 101–109.

Buchanan, C.L., Connaris, H., Danson, M.J., Reeve, C.D., and Hough, D.W. (1999). An extremely thermostable aldolase from *Sulfolobus solfataricus* with specificity for non-phosphorylated substrates. Biochem. J. *343*, 563–570.

Chaban, B., Ng, S.Y.M., and Jarrell, K.F. (2006). Archaeal habitats – from the extreme to the ordinary. Can. J. Microbiol. *52*, 73–116.

Daniel, R.M., and Cowan, D.A. (2000). Biomolecular stability and life at high temperatures. Cell. Mol. Life Sci. *57*, 250–264.

DeLong, E. (1998a). Archael means and extremes. Science *280*, 542–543.

De Long, E.F. (1998b). Everything in moderation: Archaea as 'non-extremophiles'. Curr. Opin. Genet. Dev. *8*, 649–654.

De Long, E.F., and Pace, N.R. (2001). Environmental diversity of bacteria and archaea. Syst. Biol. *50*, 470–478.

De Rosa, M., Gambacorta, A., Nicolaus, B., Giardina, P., Poerio, E., and Buonocore, V. (1984). Glucose metabolism in the extreme thermoacidophilic archaebacterium *Sulfolobus solfataricus*. Biochem. J. *224*, 407–414.

Dello Russo, A., Rullo, R., Masullo, M., Ianniciello, G., Arcari, P., and Bocchini, V. (1995). Glyceraldehyde-3-phosphate dehydrogenase in the hyperthermophilic archaeon *Sulfolobus solfataricus*: Characterization and significance in glucose metabolism. Biochem. Mol. Biol. Int. *36*, 123–135.

Deng, L., Zhu, H., Chen, Z., Liang, Y.X., and She, Q. (2009). Unmarked gene deletion and host-vector system for the hyperthermophilic crenarchaeon *Sulfolobus islandicus*. Extremophiles *13*, 735–746.

Elbein, A.D., Pan, Y.T., Pastuszak, I., and Carroll, D. (2003). New insights on trehalose: a multifunctional molecule. Glycobiology *13*, 17R-27R.

Empadinhas, N., and Da Costa, M.S. (2006). Diversity and biosynthesis of compatible solutes in hyper/thermophiles. Int. Microbiol. *9*, 199–206.

Ettema, T.J.G., Ahmed, H., Geerling, A.C.M., Van Der Oost, J., and Siebers, B. (2008). The non-phosphorylating glyceraldehyde-3-phosphate dehydrogenase (GAPN) of *Sulfolobus solfataricus*: A key-enzyme of the semi-phosphorylative branch of the Entner–Doudoroff pathway. Extremophiles *12*, 75–88.

Fabry, S., and Hensel, R. (1987). Purification and characterization of D-glyceraldehyde-3-phosphate dehydrogenase from the thermophilic arachaebacterium *Methanothermus fervidus*. Eur. J. Biochem. *165*, 147–155.

Grogan, D.W. (1989). Phenotypic characterization of the archaebacterial genus *Sulfolobus*: comparison of five wild-type strains. J. Bacteriol. *171*, 6710–6719.

Haferkamp, P., Kutschki, S., Treichel, J., Hemeda, H., Sewczyk, K., Hoffmann, D., Zaparty, M., and Siebers, B. (2011). An additional glucose dehydrogenase from *Sulfolobus solfataricus*: fine-tuning of sugar degradation? Biochem. Soc. Trans. *39*, 77–81.

Hansen, T., Wendorff, D., and Schönheit, P. (2004). Bifunctional phosphoglucose/phosphomannose isomerases from the archaea *Aeropyrum pernix* and *Thermoplasma acidophilum* constitute a novel enzyme family within the phosphoglucose isomerase superfamily. J. Biol. Chem. *279*, 2262–2272.

Hess, D., Kruger, K., Knappik, A., Palm, P., and Hensel, R. (1995). Dimeric 3-phosphoglycerate kinases from hyperthermophilic Archaea. Cloning, sequencing and expression of the 3-phosphoglycerate kinase gene of *Pyrococcus woesei* in *Escherichia coli* and characterization of the protein. Structural and functional comparison with the 3-phosphoglycerate kinase of *Methanothermus fervidus*. Eur. J. Biochem. *233*, 227–237.

Itoh, T., Yoshikawa, N., and Takashina, T. (2007). *Thermogymnomonas acidicola* gen. nov., sp. nov., a novel thermoacidophilic, cell wall-less archaeon in the order Thermoplasmatales, isolated from a solfataric soil in Hakone, Japan. Int. J. Syst. Evol. Microbiol. *57*, 2557–2561.

Jones, C.E., Fleming, T.M., Cowan, D.A., Littlechild, J.A., and Piper, P.W. (1995). The phosphoglycerate kinase and glyceraldehyde-3-phosphate dehydrogenase genes from the thermophilic archaeon *Sulfolobus solfataricus* overlap by 8-bp. Isolation, sequencing of the genes and expression in *Escherichia coli*. Eur. J. Biochem. *233*, 800–808.

Kardinahl, S., Schmidt, C.L., Hansen, T., Anemüller, S., Petersen, A., and Schäfer, G. (1999). The strict molybdate-dependence of glucose-degradation by the thermoacidophile *Sulfolobus acidocaldarius* reveals the first crenarchaeotic molybdenum containing enzyme. An aldehyde oxidoreductase. Eur. J. Biochem. *260*, 540–548.

Kawashima, T., Amano, N., Koike, H., Makino, S.I., Higuchi, S., Kawashima-Ohya, Y., Watanabe, K., Yamazaki, M., Kanehori, K., Kawamoto, T., *et al.* (2000). Archaeal adaptation to higher temperatures revealed by genomic sequence of *Thermoplasma volcanium*. Proc. Nat. Acad. Sci. U.S.A. *97*, 14257–14262.

Kim, S., and Lee, S.B. (2005). Identification and characterization of *Sulfolobus solfataricus* D-gluconate dehydratase: A key enzyme in the non-phosphorylated Entner–Doudoroff pathway. Biochem. J. *387*, 271–280.

König, H., Kralik, R., and Kandler, O. (1982). Structure and modifications of the pseudomurein in Methanobacteriales. In. Zbl. Bakt. Hyg. I Abt. Orig. Reihe. C *3*, 179–191.

König, H., Kandler, O., Jensen, M., and Rietschel, E.T. (1983). The primary structure of the glycan moiety of pseudomurein from *Methanobacterium thermoautotrophicum*. Hoppe-Seyler's Zeit. Physiol. Chemie *364*, 627–636.

König, H., Rachel, R., and Claus, H. (2007). Proteinaceous surface layers of Archaea: ultrastructure and biochemistry. In Archaea: Molecular and Cellular Biology, Cavicchioli, R., ed. (ASM Press, Washington, DC), pp. 315–340.

Lamble, H.J., Heyer, N.I., Bull, S.D., Hough, D.W., and Danson, M.J. (2003). Metabolic pathway promiscuity in the archaeon *Sulfolobus solfataricus* revealed by studies on glucose dehydrogenase and 2-keto-3-deoxygluconate aldolase. J. Biol. Chem. *278*, 34066–34072.

Lamble, H.J., Milburn, C.C., Taylor, G.L., Hough, D.W., and Danson, M.J. (2004). Gluconate dehydratase from the promiscuous Entner–Doudoroff pathway in *Sulfolobus solfataricus*. FEBS Lett. *576*, 133–136.

Lamble, H.J., Theodossis, A., Milburn, C.C., Taylor, G.L., Bull, S.D., Hough, D.W., and Danson, M.J. (2005). Promiscuity in the part-phosphorylative Entner-Doudoroff pathway of the archaeon *Sulfolobus solfataricus*. FEBS Lett. *579*, 6865–6869.

Littlechild, J.A., and Isupov, M. (2001). Glyceraldehyde-3-phosphate dehydrogenase from *Sulfolobus solfataricus*. In Methods in Enzymology, eds (Academic Press Inc, San Diego), pp. 105–117.

Lorentzen, E., Pohl, E., Zwart, P., Stark, A., Russell, R.B., Knura, T., Hensel, R., and Siebers, B. (2003). Crystal structure of an Archaeal class I Aldolase and the evolution of $(\beta\alpha)_8$ barrel proteins. J. Biol. Chem. *278*, 47253–47260.

Lorentzen, E., Hensel, R., Knura, T., Ahmed, H., and Pohl, E. (2004). Structural basis of allosteric regulation and substrate specificity of the non-phosphorylating glyceraldehyde 3-phosphate dehydrogenase from *Thermoproteus tenax*. J. Mol. Biol. *341*, 815–828.

Lorentzen, E., Siebers, B., Hensel, R., and Pohl, E. (2005). Mechanism of the Schiff base forming fructose-1,6-bisphosphate aldolase; structural analysis of reaction intermediates. Biochemistry *44*, 4222–4229.

Mukund, S., and Adams, M.W.W. (1995). Glyceraldehyde-3-phosphate ferredoxin oxidoreductase, a novel tungsten- containing enzyme with a potential glycolytic role in the hyperthermophilic archaeon *Pyrococcus furiosus*. J. Biol. Chem. *270*, 8389–8392.

Olivier, B.G., and Snoep, J.L. (2004). Web-based kinetic modelling using JWS Online. Bioinformatics *20*, 2143–2144.

Orita, I., Yurimoto, H., Hirai, R., Kawarabayasi, Y., Sakai, Y., and Kato, N. (2005). The archaeon *Pyrococcus horikoshii* possesses a bifunctional enzyme for formaldehyde fixation via the ribulose monophosphate pathway. J. Bacteriol. *187*, 3636–3642.

Ronimus, R.S., and Morgan, H.W. (2003). Distribution and phylogenies of enzymes of the Embden–Meyerhof–Parnas pathway from archaea and hyperthermophilic bacteria support a gluconeogenic origin of metabolism. Archaea *1*, 199–221.

Say, R.F., and Fuchs, G. (2010). Fructose 1,6-bisphosphate aldolase/phosphatase may be an ancestral gluconeogenic enzyme. Nature *464*, 1077–1081.

Schäfer, T., and Schönheit, P. (1993). Gluconeogenesis from pyruvate in the hyperthermophilic archaeon *Pyrococcus furiosus*: Involvement of reactions of the Embden–Meyerhof pathway. Arch. Microbiol. *159*, 354–356.

Schramm, A., Siebers, B., Tjaden, B., Brinkmann, H., and Hensel, R. (2000). Pyruvate kinase of the hyperthermophilic crenarchaeote *Thermoproteus tenax*: Physiological role and phylogenetic aspects. J. Bacteriol. *182*, 2001–2009.

Schut, G.J., Brehm, S.D., Datta, S., and Adams, M.W.W. (2003). Whole-genome DNA microarray analysis of a hyperthermophile and an archaeon: *Pyrococcus furiosus* grown on carbohydrates or peptides. J. Bacteriol. *185*, 3935–3947.

Selig, M., Xavier, K.B., Santos, H., and Schönheit, P. (1997). Comparative analysis of Embden–Meyerhof and Entner–Doudoroff glycolytic pathways in hyperthermophilic archaea and the bacterium *Thermotoga*. Arch. Microbiol. *167*, 217–232.

Siebers, B., and Schönheit, P. (2005). Unusual pathways and enzymes of central carbohydrate metabolism in Archaea. Curr. Opin. Microbiol. *8*, 695–705.

Siebers, B., Brinkmann, H., Dörr, C., Tjaden, B., Lilie, H., van der Oost, J., and Verhees, C.H. (2001). Archaeal fructose-1,6-bisphosphate aldolases constitute a new family of archaeal type class I aldolase. J. Biol. Chem. *276*, 28710–28718.

Stetter, K.O. (1999). Extremophiles and their adaptation to hot environments. FEBS Lett. *452*, 22–25.

Teusink, B., Passarge, J., Reijnga, C. A., Esgalhado, E., van der Weijden, C. C., Schepper, M., Walsh, M. C., Bakker, B. M., Van Dam, K., Westerhoff, H. V. *et al.* (2000). Can yeast glycolysis be understood in terms of *in vitro* kinetics of the constituent enzymes? Testing biochemistry. Eur. J. Biochem. *267*, 5313–29

Thauer, R.K., Kaster, A.K., Seedorf, H., Buckel, W., and Hedderich, R. (2008). Methanogenic archaea: Ecologically relevant differences in energy conservation. Nat. Rev. Microbiol. *6*, 579–591.

Van Der Oost, J., and Siebers, B. (2007). The glycolytic pathways of Archaea: evolution by tinkering. In Archaea: Evolution, Physiology and Molecular Biology, Garrett, R.A., and Klenk, H.-P., eds (Blackwell Publishing, Singapore), pp. 247–259.

Van Der Oost, J., Schut, G., Kengen, S.W.M., Hagen, W.R., Thomm, M., and De Vos, W.M. (1998). The ferredoxin-dependent conversion of glyceraldehyde-3-phosphate in the hyperthermophilic archaeon *Pyrococcus furiosus* represents a novel site of glycolytic regulation. J. Biol. Chem. *273*, 28149–28154.

Verhees, C.H., Kengen, S.W.M., Tuininga, J.E., Schut, G.J., Adams, M.W.W., De Vos, W.M., and Van der Oost, J. (2004). Erratum: The unique features of glycolytic pathways in Archaea. Biochem. J. *375*, 231–246.

Wagner, M., Berkner, S., Ajon, M., Driessen, A.J.M., Lipps, G., and Albers, S.V. (2009). Expanding and understanding the genetic toolbox of the hyperthermophilic genus *Sulfolobus*. Biochem. Soc. Trans. *37*, 97–101.

Woese, C.R., and Fox, G.E. (1977). Phylogenetic structure of the prokaryotic domain: the primary kingdoms. Proc. Nat. Acad. Sci. U.S.A. *74*, 5088–5090.

Worthington, P., Hoang, V., Perez-Pomares, F., and Blum, P. (2003). Targeted disruption of the α-amylase gene in the hyperthermophilic archaeon *Sulfolobus solfataricus*. J. Bacteriol. *185*, 482–488.

Zaparty, M., and Siebers, B. (2011). Physiology, metabolism and enzymology of Thermoacidophiles. In Extremophiles Handbook, Horikoshi, K., Antranikian, G., Bull, A.T., Robb, F.T., and Stetter, K.O., eds (Springer, Tokyo), pp. 602–639.

Zaparty, M., Esser, D., Gertig, S., Haferkamp, P., Kouril, T., Manica, A., Pham, T.K., Reimann, J., Schreiber, K., Sierocinski, P., *et al.* (2010). "Hot standards" for the thermoacidophilic archaeon *Sulfolobus solfataricus*. Extremophiles *14*, 119–142.

Zillig, W., Stetter, K.O., and Wunderl, S. (1980). The *Sulfolobus*-'Caldariella' group: taxonomy on the basis of the structure of DNA-dependent RNA polymerases. Arch. Microbiol. *125*, 259–269.

Zwickl, P., Fabry, S., Bogedain, C., Haas, A., and Hensel, R. (1990). Glyceraldehyde-3-phosphate dehydrogenase from the hyperthermophilic archaebacterium *Pyrococcus woesei*: Characterization of the enzyme, cloning and sequencing of the gene, and expression in *Escherichia coli*. J. Bacteriol. *172*, 4329–4338.

Index

www.ingramcontent.com/pod-product-compliance
Lightning Source LLC
Chambersburg PA
CBHW071257180926
51613CB00001B/49
* 9 7 8 1 9 0 8 2 3 0 0 2 7 *